W9-CSP-035

The Design and Application
of Process
Analyzer Systems

CHEMICAL ANALYSIS

A SERIES OF MONOGRAPHS ON
ANALYTICAL CHEMISTRY AND ITS APPLICATIONS

VOLUME 70

A WILEY-INTERSCIENCE PUBLICATION

JOHN WILEY & SONS

New York / Chichester / Brisbane / Toronto / Singapore

The Design and Application of Process Analyzer Systems

PAUL E. MIX

Corpus Christi, Texas

A WILEY-INTERSCIENCE PUBLICATION

JOHN WILEY & SONS

New York / Chichester / Brisbane / Toronto / Singapore

Copyright © 1984 by John Wiley & Sons, Inc.

Library of Congress Cataloging in Publication Data:

Mix, Paul E.
    Design and application of process analyzer systems.

    (Chemical analysis, ISSN 0069-2883 ; v. 70)
    "A Wiley-Interscience Publication."
    Includes index.
    1. Chemical process control—Equipment and supplies.

I. Title. II. Series.
TP155.75.M58 1984   660.2'81   83-21915
ISBN 0-471-86518-4

Printed in the United States of America

10 9 8 7 6 5 4 3 2 1

PREFACE

The purpose of this book is to provide basic guidelines for the design of process analyzer systems. Each component in the analyzer system is examined in detail. Critical design information is provided when needed, and general design philosophy has also been included.

The on-line analyzer can be regarded as a specialized process control or process monitoring instrument. The sample system of the analyzer should also be considered an integral part of the total system design.

However, problems will still be encountered even when on-line analyzers are properly applied and sample system designs are adequate. The reason for this is that on-line analyzers are frequently used to monitor wet, dirty, or corrosive process streams. When troubles do arise, the problems will often be relevant to the sample system as opposed to the analyzer hardware. This is primarily because modern technology has greatly increased the reliability of electronic circuits and mechanical devices now used in modern on-line process analyzers.

On-line analyzer results are often compared against analytical laboratory results. Although the basic measurement principles may be the same, the instrumentation used is often quite different. Laboratory instruments are more versatile, accurate, and delicate. On-line analyzers are often dedicated to a single measurement and are more rugged.

This book discusses process analyzer applications and some of their common problems and offers guides to help the reader find solutions. Production, maintenance, or technical personnel concerned with the application of process analyzer systems should find this information helpful. Instrument engineers especially should find this book to be a handy reference.

PAUL E. MIX

Corpus Christi, Texas
March 1984

v

ACKNOWLEDGMENTS

The author would like to extend a special note of thanks to Duane Poland, designer–draftsman, for his original artwork and to W. C. Welz, Jr., analyzer specialist, for editing the original manuscript and writing the chapter on gas chromatography.

The author would also like to thank the numerous instrument and sample system component manufacturers listed in Appendix 2 and throughout the text for supplying pertinent application notes, sketches, catalogs, and instrument manuals for review prior to the publication of this work.

P.E.M.

CONTENTS

ix

The Design and Application of Process Analyzer Systems

CHAPTER

1

INTRODUCTION

On-line analyzers are field-mounted analytical instruments designed for continuous use in almost any process or manufacturing industry. They measure one or more process stream components at one or more process locations. On-line analyzers can be described as "eyes of the process" and can be designed to:

1. Monitor an electrical characteristic of the process stream such as pH or conductivity.
2. Measure the concentration of an element or chemical compound in the sample.
3. Alleviate some of the work load associated with analyzing large numbers of routine samples.
4. Operate as an active element in an instrument control loop.
5. Provide an input to a computer system for subsequent supervisory or control functions.
6. Protect personnel and equipment from potential hazards.

Some of these functions may overlap; however, the list is certainly not all-inclusive. On-line analyzers are thus an important consideration during the initial design phases of a new plant or manufacturing process.

At the present time, the state of the art in the development of process analyzers is about one generation behind the development of laboratory instruments and two or three generations behind defense electronics. Nevertheless, an increasing number of manufacturers are rapidly incorporating microprocessors into their designs. Within 10 years an on-line analyzer without a microprocessor will probably be a rarity.

Microprocessors are being used to correct zero baseline and span drift; characterize pH signals to match process titration curves; and perform self-diagnostics on sensors, electronic units, and interconnected accessories. In addition, they can perform calculations, expand ranges, program alarm setpoints, and provide control outputs. In some multiple sensor installations microprocessors provide sequencing or track-and-hold functions.

Most microprocessor-controlled units feature user access through a front panel keyboard and provide prompting routines to prevent programming mistakes. Microprocessors have greatly increased the versatility and usability of on-line analyzer systems.

Some analyzers currently using microprocessors include pH monitors, mass spectrometers, digital titrators, calorimeters, specific ion analyzers, hygrometers, and various gas analyzers including chromatographs. Even though the use of microprocessors is rapidly increasing, analyzers are still dependent on their sample systems for a representative sample of process material. Therefore, one must consider the total analytical system from sample source to analysis display when evaluating the applicability of a particular analyzer design.

1.1. THE TOTAL SYSTEM CONCEPT

Electronic sophistication cannot compensate for a poorly designed sample system. The purpose of the sample system is to deliver a representative sample of the process material to the on-line analyzer. Residence time of the sample system may be of prime importance if the analyzer output is used in process control. For an on-line analyzer system to be successfully applied to a process, equal importance should be placed on sample system design. Figure 1.1 shows that an on-line analyzer system consists of a sample preconditioner unit, sample system, and the analyzer hardware.

If the response time of the process is rapid, sample lines should be kept as short as possible and sample preconditioners should be mounted as close to the process tap as practical. A typical sample preconditioner consists of a "roughing" filter, sample coolers or heaters, pressure regulating equipment, and valves for isolating or venting the components. Preconditioning prepares the sample for sample system transport to the on-line analyzer. This may involve removing suspended solids that might plug sample lines or other downstream components, changing the physical state of the sample, or intercepting contaminants that might damage the analyzer or degrade its analysis.

After the process sample has been preconditioned, it passes on to the analyzer by way of the sample system. The sample system is usually installed in a manner to bring the preconditioned sample to the analyzer by way of the fastest route. A typical sample system usually consists of valves, switchable filters, pressure gauges, pressure or flow control devices, and often pressure-relief devices or reverse-flow check valves. Also, buffering chemicals or other agents are added to the sample at this

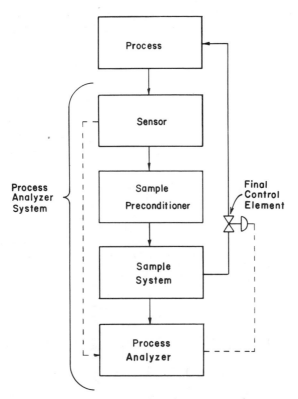

Figure 1.1. Block diagram representation of a process analyzer system.

point to enhance the performance of the process analyzer. Sample systems are covered in detail in Chapter 2.

Process analyzers that directly or indirectly control a final control element are considered to be in "closed-loop control." Process analyzers are in direct control when their output goes to a controller that uses the analysis result as the process variable. Typical of these are conductivity and pH monitors, hygrometers, specific ion analyzers, and single-component gas analyzers. The analyzer is in indirect control when its output goes to a computational device that performs calculations on the basis of the analysis result. Calorimeters for total British thermal unit (BTU) determinations, densitometers, or gravitometers, and most gas–liquid chromatographs fit into this category. Both direct and indirect control schemes are shown in Figure 1.2.

Cost is definitely an important consideration for process analyzer system design. Analyzers cost $1500 to $25,000 each (if mass spectrometers

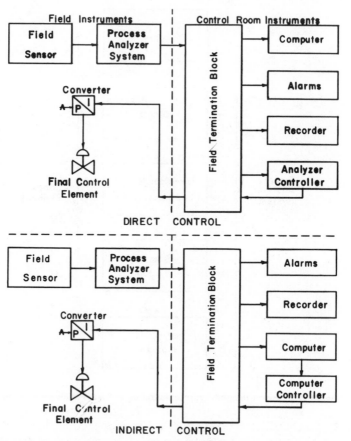

Figure 1.2. Example of direct and indirect control using process analyzers. With direct control, the analyzer controller directly controls the final control element. With indirect control, the computer controller controls the final control element.

are considered, upper cost could be as high as $100,000), and the old idea that one could total the hardware costs and multiply by 2 is no longer viable when it comes to project cost estimation. The analyzer may be relatively inexpensive and its sample system expensive or vice versa. A treatise on project estimating would be too lengthy for inclusion in this book; however, a few considerations of cost estimation are included as guidelines. Estimates vary considerably with geographic location, plant size, labor market, and of course, process requirements. A few of the major cost factors to be considered are:

1.   Materials, based on recent cost quotations. In most cases vendor quotations are valid for only 30 days as a result of inflation.

2. Labor, based on days worked, hourly rate, and average worker efficiency.
3. Design, typically 10 to 15% of total project cost. Contract engineering, drafting, or blueprint services would be included here.
4. Freight, based on mode of transportation, distance, and rates.
5. Escalation, typically 0.1% per month for labor and material. The extent to which costs are likely to increase from the start of your project to its finish must be determined.
6. Overhead, typically 50 to 60% of labor costs. Supervisory costs and the continuing costs of running a business, such as rent and maintenance, are included.
7. Undeveloped project scope, typically 10% of total project cost. This is used to cover miscellaneous materials and labor costs that were not firm at the time of the original project estimate.

The cost factors previously mentioned are not intended to be all-inclusive but are included here to provide the reader with the more volatile factors that influence project cost estimation. The factors considered must be tailored to each estimator's work environment. Cost estimation requires careful consideration and attention to details. The profitability of a project and its return on investment depend greatly on accurate cost estimation.

When a manufacturing plant or process is being designed, a careful selection of on-line analyzer systems should be made. Analyzers are like the U.S. marines; a few good analyzers are usually all that are really needed. On-line analyzers should not be installed because "it would be nice to have them" or because there is enough "fat" in the project to pay for them. Such reasoning results in premature obsolescence.

Production technicians ignore analyzer systems that are not really needed. If production ignores them, maintenance technicians have no incentive to maintain them and keep them running. Furthermore, as rugged as they are, no instrument will operate unattended indefinitely. Eventually the system fails and is salvaged or sold for scrap.

The most successful analyzer applications are those where the analyzers are used in closed-loop control or are used to interlock the production process when an unsafe operating condition exists.

1.2. EQUIPMENT PROTECTION

On-line analyzer systems are costly and should be adequately protected from the environment and the process streams that they are monitoring.

In the Gulf Coast area, heat, humidity, and salt spray are factors to be considered. The cost of a single preconditioner can run as high as $10,000 when fabricated from materials required for environmental protection. Painted or galvanized steel has been shown to have a relatively short life in the Gulf Coast environment. As a result, stainless steel boxes or fiberglass boxes with stainless steel hardware have often been used.

Some instrument engineers may not consider simple pH or conductivity meters as in-line analyzers. However, they are in the true sense of the word because their sensors are usually placed directly in the sample stream. In addition, they are frequently used to monitor continuous operations and are often placed in closed-loop control. These instruments can be mounted on pipe stands in the field because their sample systems usually consist of only a few valves and little or no sample preconditioning equipment. Even when this is the case, to avoid damage, their front panels must not be exposed to direct sunlight. Likewise, analyzers should not be mounted under process equipment where chemical spills might damage them.

In more complex systems, the preconditioner is mounted in the field near the process tap. From there the sample travels to the main sample system that may be mounted in one room of a multiroom analyzer house. The analyzer hardware is often installed in an adjacent room separated from the sample system by fume barriers. This electronics room can be air-conditioned to cool equipment and remove harmful humidity.

Analyzer protection can vary from its simplest form such as a National Electrical Manufacturer's Association (NEMA) 4 case mounted on a pipe stand, to more sophisticated protection where the analyzer and its sample system are mounted in an outdoor fiberglass or stainless steel enclosure. Some applications require that a number of analyzers and their sample systems be mounted in a multiroom analyzer house complete with ventilation and ambient-air monitoring for detection of hazardous fumes or dangerous gases. The method of protection provided depends on the basic design of the analyzer and the complexity of its sample system and the number of analyzers to be protected in a given plant area.

1.3. INTERACTION WITH OTHER LOOP COMPONENTS

Whenever an on-line analyzer malfunctions, the first statement usually made is, "Something's wrong with the analyzer." In reality, this statement may mean that the sample system is plugged, the analyzer results disagree with the analytical laboratory results, or the computer or panel instruments disagree with the analyzer display.

When problems arise, the basic sensor may also be at fault. Are the pH probes responding to buffer solutions? Are optical cells clean and free of dirt or films? Are the sensors actually being delivered a representative sample of the process material? These and many other questions need to be asked before maintenance technicians rush into detailed analyzer troubleshooting.

A detailed electronic troubleshooting procedure is usually the final step in a systematic investigation of the problem. A little forethought and planning will often save hours of misdirected effort to resolve analyzer "problems." Additional guidelines on troubleshooting are given in Chapter 10.

CHAPTER

2

BASIC SAMPLE SYSTEMS

In its simplest form, a pH meter or dissolved oxygen analyzer may have its sensor immersed in a plant effluent stream. This is a common practice with regard to pollution monitoring equipment required by governmental pollution control agencies.

Other sensors that lend themselves to this sort of direct application include conductivity probes, hygrometer probes, and in some cases specific ion probes. In each case, flow of the stream must be sufficient to assure a reliable measurement.

When sample system requirements are more complex than this, a direct tap into the process pipeline may be needed. Small-diameter sample system lines are often connected to larger-diameter process lines by means of blind flanges that have been drilled and tapped to accept a standard tubing fitting. Generally a full-port ball valve is located immediately downstream of the sample tap so that the sample system can be blocked for repairs. Welded construction and small-diameter pipes can also be used for sample system lines.

In the petrochemical industry, it is desirable to restrict the flow and volume of sample brought inside the analyzer house. This can be accomplished by a restricting orifice, porous micron (sintered metal) filter, or small metering valve in the sample system. The purpose of the device is to reduce the flow and quantity of flammable, explosive, toxic, or corrosive material in the event of a sample line break or rupture.

Figure 2.1(a) through (d) shows various sample tap arrangements. Figure 2.1(a) shows the sample system tubing connected to a downward-facing tee. This is a particularly poor arrangement because solids and dirt are more likely to be swept along the bottom of the pipeline and be pulled into the sample system. Figure 2.1(b) shows a top tap, which is also considered a poor practice. A top tap is adversely affected by laminar flow, gas bubbles, and lighter-density gases. The side tap shown in Figure 2.1(c) is a better choice because a more representative sample of liquid or gas should be present near the vertical centerline of the pipe. The best low-pressure sample tap configuration shown in Figure 2.1(d) is one where the sample tubing extends through the pipe tee into the center of

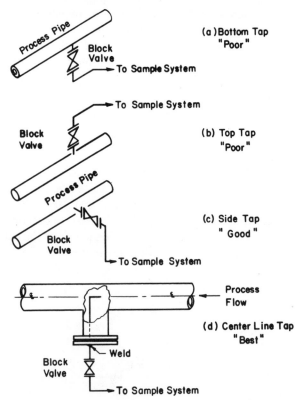

Figure 2.1. Various sample tap configurations with the preferred configuration shown in 2.1(d). Sample tubing should not extend more than 20% into the inner diameter of 2-in. or larger pipes.

the pipe and faces upstream against the direction of process flow. In high-pressure systems the tubing should extend less than 20% into the pipe and face downstream. With the centerline tap configuration, the pipe tee facing direction is unimportant.

Other arrangements covered in this chapter include direct mounting of analyzers on process pipes, utilization of high-speed bypass lines, and the design of vented analyzer systems where the sample stream is routed to a vent header rather than returning to the process.

2.1. PROCESS LINE INSTALLATIONS

A process line installation either routes the entire process stream through the analyzer, or the sensor can be directly inserted in the process pipeline.

Small sensors such as pH, conductivity, or moisture probes can be easily installed directly into a process stream by threading them into a blind flange. One problem with this arrangement is that it may be difficult to remove or service the sensor without shutting the process down. To circumvent this problem some manufacturers provide probe designs that allow probes to be inserted through a full-port ball valve. This type of probe usually has a mechanical stop so that the probe cannot be fully removed.

Figure 2.2 shows a direct insertion probe arrangement. The advantage of this probe arrangement is that it is mounted directly into the process; therefore, no sample system is required. The sensor must be compatible with the process and designed for use at the process pressure and temperature. The disadvantage of this arrangement is that the probe can blow out if the safety stop is not properly designed and used. In practice, the probe is withdrawn until the safety cable is tight. The length of the cable is set so that the probe can be withdrawn only to the point where the ball valve can be closed. After the ball valve is closed, one end of the safety cable is

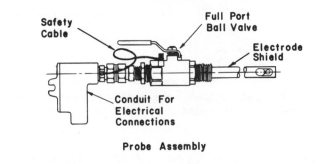

Probe Assembly

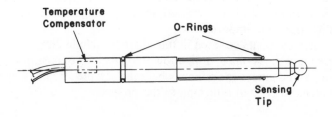

Replacement Electrode

Figure 2.2. Diagram of a pH probe designed for direct insertion into the process through a full-port ball valve. The ball valve is closed before final probe removal and safety cable release. Courtesy of Van London Company.

disconnected and the probe is completely removed. The O rings on the probe act as process seals until the ball valve can be closed.

Some optical analyzers can be mounted directly on the process pipe or stack. To accomplish this, a pipe section containing transparent windows is installed in the line. Usually the energy source is placed on one side of the line and the detector is placed on the other. However, if source and detector are on the same side, a mirror is placed on the other side and the energy is reflected. The window-to-window dimensions or twice the window-to-mirror distance is the effective light pathlength of the cell. Figure 2.3 shows various source and detector arrangements. Other than the special cell windows, these arrangements require few, if any, other sample system parts. They also have the distinct advantage that nothing is inserted in the sample stream. Chances of leaks or blowouts are greatly

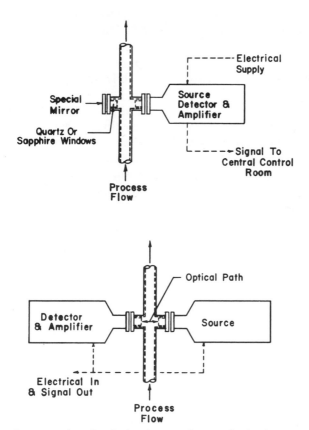

Figure 2.3. Direct mounting of optical analyzer such as smoke densitometer on a process pipe. A process shutdown is required for optics cleaning.

reduced provided window seals are compatible with the physical parameters of the sample stream.

There is one important disadvantage with this type of system: keeping sample cell windows clean. Sometimes a cleaning agent such as steam can be backflushed through the cell periodically to keep the windows clean. Of course, this requires that the process be compatible with water.

It should be emphasized again that it may be difficult to maintain the analyzer sensor or cell if it is inserted into or is installed as an integral part of the process pipeline.

2.2. BYPASS LINE INSTALLATIONS

Installation of an analyzer cell or sensor directly on a high-speed bypass line has the same advantages mentioned is Section 2.1 and fewer disadvantages. A bypass line is usually installed around a valve, pump, compressor, orifice, or some other device where a suitable differential pressure exists to overcome the combined pressure drops of all devices in the sample flow path. In most cases a small differential pressure is adequate provided flow through the bypass line is high, sample residence time is short, and the system is "fast-acting." It is also easy to add block, bleed, or bypass valves to the bypass line installation so that the system can be maintained without affecting plant operation.

Figure 2.4 shows a typical bypass line installation with suitable valving and pressure gauge monitoring. The pressure gauges are used to troubleshoot the system. During normal operation the bypass and drain valves are closed and the block valves are open. The pressure drop across the orifice should be sufficient to assure good flow through the sample system. Normally a differential of only a few pounds would be observed between the upstream and downstream pressure gauges. For maintenance of the analyzer, the block valves nearest the process pipe would be closed first. Then the bypass and vent valves would be opened, at which time both gauges should be observed to vent to a lower pressure. After this all other valves in the system could be closed.

To bypass the analyzer during normal process running conditions, the block valves nearest the analyzer cell would be closed and the bypass valve opened. It would still be unsafe to work on the system until the sample cell line is vented down.

One aspect of sample system design that should be considered is the accessibility of sample system parts and components. The location of sample taps, block valves, and all other components should be within safe reach. Sample system devices located under analyzer houses or in over-

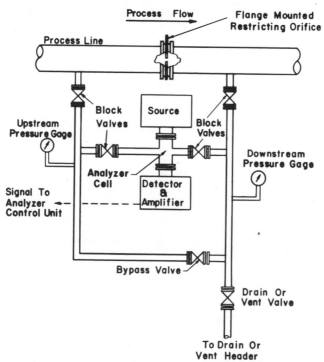

Figure 2.4. Photometric analyzer mounted on a bypass line. Sample flow can be established by differential pressure across a valve, pump, or orifice plate.

head pipe racks complicate maintenance, increase costs, and introduce unnecessary risk factors into otherwise routing operating or maintenance procedures. These devices should be located at ground level or be within easy reach from permanent platforms.

When sample flow requirements are small and the distance from the process line to the desired analyzer location is great, a bypass line and vented sample system similar to that shown in Figure 2.5 is sometimes used. The high-speed bypass line across the pump provides rapid sample flow past the preconditioner tap and keeps the high-volume, high-velocity process material in a piping loop away from the analyzer.

Analyzer flow requirements are usually low compared to the total bypass flow. The pressure regulation devices at the preconditioner can be used to control pressure while also providing adequate flow through the sample line. If the analyzer flow requirement is insignificant compared to the total flow, the sample can be disposed of instead of being returned to the process. This method assumes that the cost of a small sample loss is

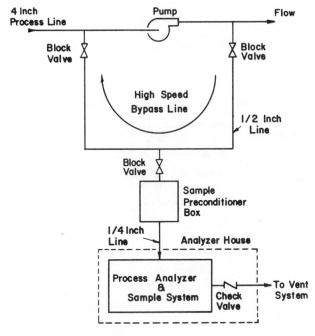

Figure 2.5. High-speed bypass line with vented sample. It may be more economical to dispose of the sample than return it to the process.

also insignificant compared to the cost of returning the sample to another point of lower pressure in the system.

One reason for not returning the sample would be the installation and maintenance costs of a long sample return line, especially if steam or electrical heat tracing were required. Maintenance of the line can be aggravated if the sample fluid is corrosive or likely to cause pluggage.

In gas sampling systems, the exit gas from the analyzer can be vented into a large collector or header system. This gas can then be treated to remove environmentally unacceptable compounds before being discharged. In liquid sample systems the analyzer effluent may drain into a waste hold tank or other receiver where it can be treated to remove pollutants before being discharged.

A check valve isolates the sample system components from the vent system and prevents backup of vent header vapors into the sample system. The vent headers can sometimes have a slight vacuum and at other times have a small positive pressure depending on vent rates from other collection points along the system.

2.3. SAMPLE SYSTEM OPERATION

Figure 2.6 shows a typical gas sample system used successfully in corrosive service applications. A gas system has been chosen for purposes of discussion since liquid systems are usually less complicated. Comments pertinent to liquid sample systems are covered as individual sample system components are discussed.

Some of the significant features of this system are:

1. Dual sample filters with isolation block valves allowing filter changes without disrupting sample flow to the analyzer.
2. Automatic temperature and pressure control to help maintain analyzer calibration integrity.
3. Permanent tube connections for introduction of calibration gas or plant utilities to the system to simplify maintenance and reduce tubing stress and tube fitting wear.
4. Pressure and flow indicating devices installed at locations where

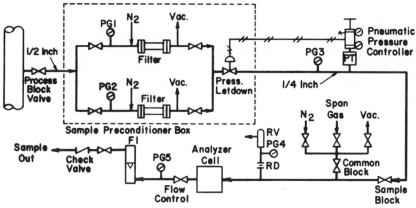

Notes :
1) Steam Tracing Not Shown
2) All Valves Not Shown
3) Component Identification For
   Purposes Of Discussion

Figure 2.6. A typical gas analyzer system incorporating many of the design features discussed in Chapter 2.

important system operating parameters must be monitored for effective system maintenance.

5. The use of special control devices such as rupture disks, relief valves, flow limiters, and check valves to protect personnel and equipment.

6. All tubing and other sample system components made from materials that exceed the corrosion resistance rating of the process pipe material.

2.3.1. Sample Preconditioning

Sample preconditioning prepares the sampled fluid for introduction into the sample transport system. Samples may require straining, heat exchange, filtering, or pressure control after removal from the process pipe.[1,2]

Large amounts of pipe scale and other contaminants can be encountered during new plant commissioning startups. This is usually caused by inadequate pipe flushing procedures during the first phases of plant operation. Some of these contaminants may be dust from column packing materials, trash left in the pipes during construction, or oils from machining tools used during pipe fabrication.

It is also common practice to reduce sample tube size after the sample has been preconditioned to maximize sample flow velocity and minimize sample residence time in the tubing.

Dual in-line filters are often used so that sample flow is not interrupted if one filter must be removed from service. The isolation valves must be installed with enough tubing clearance to allow filter removal without unduly stressing the tube connections on the other filter. If the system is too rigid, the strain of pulling or pushing the tubing might eventually cause leaks at the other tube fittings.

Frequent filter maintenance may be required until the plant has been in operation long enough for the piping systems to become free of contaminants. Daily filter maintenance might initially be required; later, intervals of over 3 months might be adequate. One should never loose sight of the fact that the purpose of the filters is to remove dirt; if anything in the sample system should plug it, it should be the filters!

A filter must be able to withstand the composition of the sample stream and meet the flow, temperature, and pressure requirements of the analyzer while providing effective filtration without undue maintenance. Care must also be given to the selection of core materials and filter housings to ensure chemical compatibility with sample fluid and meet system operating requirements. Reinforced fiber filters and sintered-metal filters are well suited for many analyzer applications and are widely available. Table 2.1 summarizes some fiber filter applications and lists their maximum temperature ratings.

Filters capable of retaining particles of 10 μm or less in size will usually allow gas sample flow of up to several standard cubic feet per minute with less than a 2-lb pressure drop. In liquid systems the filter pressure drop may be more critical. Figure 2.7 shows the affect of liquid flow on pressure drop with varying particle size filtration. The curves are based on nonpolar liquids with a fluid viscosity of 1 centistoke. Filters with porosities of 10 to 50 μm are often used to minimize pressure drops in liquid service. Since some fibers swell in aqueous solutions, pressure drops across them may be 4 to 5 times the drop across nonswelling fibers.

Small sintered-metal filters for $\frac{1}{8}$-, $\frac{1}{4}$-, and $\frac{1}{2}$-in. tubing are often used for sample filtration. Because of their low internal volume, they are best

Table 2.1. Fiber Filter Materials and Applications

Material	Applications	Temperature, °F
Bleached cotton	Potable liquids, dilute acids and bases, organic solvents and gases	300
Rayon-cellulose	Same as above, but fiber swells more in aqueous solutions	300
Modacrylic	Strong bases, acids, and oxidizing agents	200
Polypropylene	Same as above except unsuitable for oxidizing agents	275
Baked glass fiber	Where pure glass is recommended, high-temperature applications	750
Nylon	For concentrated bases, condensate, and organic solvents	300
Acrylic	For strong acids and organic solvents	300
Jute–rayon	For viscous fluids and paint	250
Nomex®	For steam filtration	310
Teflon®	For steam, high-temperature condensate, and extremely corrosive fluids	500

Source: Courtesy of Filterite/Brunswick Corporation.

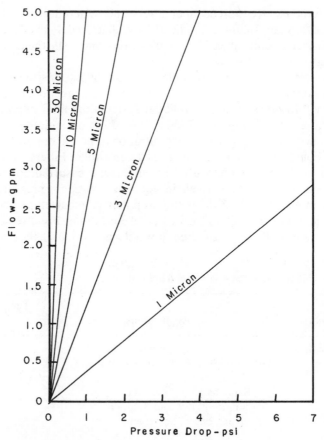

Figure 2.7. Flow–pressure drop characteristics for filters of differing micron particle size. Courtesy of Filterite/Brunswick Corporation.

applied to clean liquid or gas streams having relatively low flow rates. Small porous metal filters may also have a secondary benefit in acting as flame arrestors in sample streams handling flammable fluids. Table 2.2 shows the maximum air and water flows that can be obtained for various porosities and pressure drops. The table applies only to ¼-in. filters. Flow rates are greater for larger filters and less for smaller ones.

2.3.2. Pressure-Control Devices

Self-contained single-stage pressure regulators similar to the one shown in Appendix 7 can be used for relatively clean, dry, low-volume sample

Table 2.2.   Flow Characteristics of ¼-in. Sintered-Metal Filters

Fluid	Micron Size[a]	Maximum C_v[b]	Maximum Flow or Air (scfm) or Water (gpm) at Stated Pressure Drop to Atmosphere		
			10 psi	50 psi	100 psi
Air	2.0	0.096	1.33	3.67	6.57
Water	2.0	0.096	0.30	0.68	0.76
Air	7.0	0.15	2.08	5.74	10.17
Water	7.0	0.15	0.47	1.06	1.50
Air	15	0.21	2.93	8.03	14.24
Water	15	0.21	0.66	1.48	2.10
Air	60	0.48	6.64	18.36	32.55
Water	60	0.48	1.52	3.39	4.80
Air	90	0.50	6.92	19.13	33.91
Water	90	0.50	1.58	3.54	5.00

Source: Adapted from Nupro Company information copyright 1979, 1980, Markad Service Company. All rights reserved.
[a] Of particles this size, 90% will be blocked by the filter.
[b] The C_v factor is related to the flow through a given size orifice at a specific differential pressure and temperature.

streams. This type of pressure control device is used extensively in the chemical and petrochemical industries.

In cases where the sample may be dirty, toxic, or corrosive, a sample system similar to the one shown in Figure 2.6 may be used.[3] The pressure let-down valve is located inside the preconditioner box, but the pressure sensor and transmitter are located inside the analyzer house. The primary reason for controlling sample pressure is that many analyzers are designed for use at a specific detector pressure that is often near atmospheric pressure. If the analyzer is a gas-phase analyzer, detector calibration will depend on gas sample pressure and temperature in accordance with Beer's law. Calibration equations for ultraviolet (UV) and infrared (IR) analyzers are given in Chapter 7. Moreover, leaks are less likely to occur at lower pressures.

The sample preconditioner is located near the process sample point. However, pressure gauge PG3 and pressure transmitter PT can be located remotely in the analyzer shelter. If the process is operating at a relatively high pressure, the sample system pressure transmitter must be capable of operating anywhere in that range depending on the pressure limits of the system and the requirements of the analyzer. The sample block valve is a full-port block valve that has little or no pressure drop. The pressure gauge and transmitter, valves, and relief system are all located near the analyzer cell and operate at essentially the same pressure. The calibration gas and vacuum purge manifold is located downstream of the sample block valve so that the cell can be quickly evacuated and purged. In this system nitrogen is used to help purge the system as well as acting as a zeroing gas; therefore, it is introduced just upstream of the analyzer cell for fastest response.

When the analyzer system or plant is shut down, nitrogen is left flowing to keep the system dry and free of contaminants. Water is often used in aqueous analyzer systems as the zeroing medium and left circulating when the analyzer is not on line.

Rupture Disk and Relief Valve

Sample systems that use rupture disks for pressure protection will be safe only if the flow capacity of the disk is relatively large compared to the maximum flow capacity of the sample line. A large flow allows the disk to easily vent excessive pressure in the small-volume tubing.

As shown in figure 2.6, it is relatively easy to burst the rupture disk through improper sequential operation of the valves. For example, if any downstream valve is shut, system pressure will increase even though the pressure control valve closes. This is because metering valves are seldom capable of tight shutoff since they do not have the soft seats required to totally stop flow. They are designed primarily for throttling applications and use hard metals internally to minimize wear. Pressure gauge PG4 is used to show when the rupture disk has burst. The rupture disk in Figure 2.6 protects the gauge and relief valve above it from corrosion process material and subsequent failure to operate when required. The selection of a rupture disk and its correct sizing require that the user provide additional capacity for fast reactions and the unexpected. Rough approximations of flow and the required rupture disk area can be determined from sizing tables, nomographs, and equations. Table 2.3 lists various flows as a function of pressure and rupture disk diameter.

Table 2.3 shows that flow through the rupture disk can be relatively large compared to the sample flow using 5000 cm³/min for a typical gas sample flow rate and 2 gpm for a typical liquid sample system. A gas flow of 5000 cm³/min is about 0.18 scfm, so almost any rupture disk would be able to vent the entire sample system even if the pressure control valve were fully open. The same situation should exist with the relief valve.

Small relief valves with tubing connections are frequently used in sam-

Table 2.3. Rupture Disk Sizing Table[a]

Rupture Pressure in psig	Gas Flow in scfm as a Function of Disk Diameter in Inches			
	0.5	1.0	1.5	2.0
50	145	580	1300	2300
100	256	1030	2320	4100
200	478	1900	4300	7500
400	925	3700	8300	14800
	Liquid Flow in gpm as a Function of Disk Diameter in Inches			
50	32	125	282	500
100	45	180	400	720
200	63	250	560	1000
400	90	355	800	1420
	Dry or Saturated Steam Flow in pph as a Function of Disk Diameter in Inches			
50	390	1520	3450	6100
100	690	2700	6100	10800
200	1290	5040	11400	20200
400	2490	9700	22000	39100

Source: Adapted from Nupro Company information copyright 1973, 1981 by Markad Service Company. All rights reserved.
[a] Gas referenced to air at 14.7 psig and 60°F.

ple system designs. They usually consist of a spring-opposed poppet that seats against an elastomeric O ring. Their cracking pressure is a function of the area of the poppet exposed to upstream pressure versus the poppet spring pressure rating. Downstream pressure effects can be omitted if the relief valve opens to atmosphere. Figure 2.8 shows typical pressure drop–flow characteristics.

Table 2.4 is a partial listing of flow capacity as a function of relief valve size, design, and cracking pressure, which in this case is the same as the pressure drop to atmosphere. Note that at 10 psi a ¼-in. relief valve would not be able to relieve 2-gpm equivalent water flow. At higher relief pressures a ¼-in. valve would be adequate.

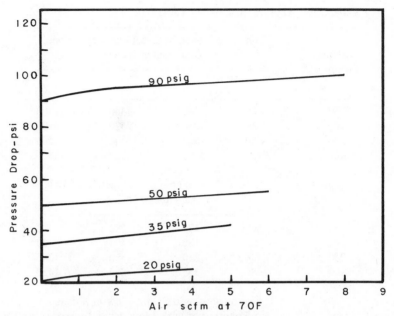

Figure 2.8. Relief valve pressure drop–airflow characteristics. Normal relief valve pressure is given for each curve. Courtesy of Nupro/Markad Service Company.

Table 2.4. Flow Capacity of Small Sample System Relief Valves

Pressure Drop to Atmosphere dP, psi	¼-in. Pipe Ended		½-in. Pipe Ended	
	Air scfm at 70°F	Water gpm at 70°F	Air scfm at 70°F	Water gpm at 70°F
10	4.82	1.11	16.69	3.79
50	13.25	2.48	45.44	8.49
100	23.49	3.50	80.55	12.00
	¼-in. In-Line		½-in. In-Line	
10	6.50	1.49	22.96	5.25
50	17.98	3.32	63.49	11.74
100	31.87	4.70	112.56	16.60

Source: Adapted from Nupro Company information copyright 1973, 1981, Markad Service Company. All rights reserved.

22

Pressure Gauge Monitoring

Some sample systems must handle dirty or corrosive streams from plant processes containing corrosive products such as iron oxide and nickel chloride. In some cases the sample stream may be so corrosive that contaminants are generated within the sample system itself. Finely divided solid particles resembling a cloudy film are of great concern in IR photometric analyzer systems that depend on both direct energy and reflected energy transmission for a satisfactory signal: noise ratio.

In the apparatus shown in Figure 2.6 five pressure gauges have been installed for troubleshooting the sample system. The instrument technician can verify that the sample system is operating properly by monitoring these field gauges. Daily inspections might be required during startup or when the analyzer system is considered crucial to plant operations. This might be the case when the on-line analyzer is used to interlock or shut the plant down as a result of an unsafe operating condition or when the operating department will not run the plant without the analyzer. In less critical service a weekly or even monthly inspection might be all that is required. During startup it is better to inspect the on-line analyzer system more often than necessary until some experience is obtained with sample system performance.

The use of pressure gauges for troubleshooting can best be explained by considering what happens if line pluggage occurs at various parts of the sample system. For example, if a plug forms in the process block valve, pressures gauges PG1 or PG2, PG3, and PG5 will all bleed down to sample outlet pressure. If the filter that is in service becomes impassible, the pressure gauge associated with that filter will remain at process line pressure while gauges PG3 and PG5 will bleed down to downstream pressure. When the sample system is blocked between PG3 and PG5, PG3 will gradually build up to process line pressure but PG5 will remain at downstream pressure. If the sample system block occurs downstream of PG5, all pressure gauges in the system will build up to process line pressure. A relief system is necessary because PG3, PG5, the analyzer, and the rotameter are usually designed for lower pressures than the components upstream of the pressure control valve.

Pressure gauge PG4 will not show pressure unless the rupture disk pressure limit is exceeded. The same pressure will then appear on both PG3 and PG4. The rupture disk is usually made of exotic materials such as Inconel,® Hastelloy,® Monel,® or pure nickel to protect it from corrosion. The pressure gauge and relief valve are usually made of common materials such as mild steel or brass but do not come in contact with the sample stream until the disk ruptures.

Check Valves

Check valves are similar to on-line relief valves differing principally in their spring settings and flow capability. Small relief valves may have cracking pressures of several hundred pounds, whereas check valves often have low pressure range springs that provide pressures as low as ⅓ psi. The main purpose of a check valve is to prevent reverse flow of the sample while offering as little resistance as possible to the forward flow of gas or liquid.

Check valves are also used with valving manifolds to protect air, nitrogen, and calibration gas cylinders against contamination from the process sample. For example, the sample might be at a much higher pressure than the nitrogen purge gas supply. Improper valving could result in process gas flowing into the nitrogen system. Check valves can help prevent this if they are properly located at the header's common tie points. If several purge gas service lines are tied to a common manifold, care must be taken to prevent back-pressuring the header with the process gas from higher pressure sources tied to the other purge gas service lines.

The check valve represents a restriction in the line because it usually contains a spring-loaded poppet opposing the flow. During reverse flow conditions, additional downstream pressure assists the spring in sealing the poppet against its seat. The check valve is a primary point in the sample system where solids can accumulate, so it should be one of the first items checked when the pressure gauges indicate downstream blockage.

Back-Pressure Control

Back-pressure control is sometimes used with on-line analyzer systems to hold sensor or detector pressures constant in systems where the sample outlet is subjected to pressure variations. Single-stage manual back-pressure regulators can be used to provide adjustable back-pressure ranges of 0 to 25, 50, 100, 250, or 500 psi. They are designed for use in relatively clean gas or liquid streams such as might be encountered in many petrochemical plant applications. In addition to providing good throttling action, these regulators also provide bubbletight shutoff.

A pressure let-down valve and controller ahead of a sample cell is one way to hold sample cell pressure constant in systems where the sample outlet pressure does not change. A back-pressure regulator after the sample cell may be required to hold sample cell pressure constant if the downstream pressure varies as it does in flare headers or process return lines. Figure 2.9 shows an automatic back-pressure control system that

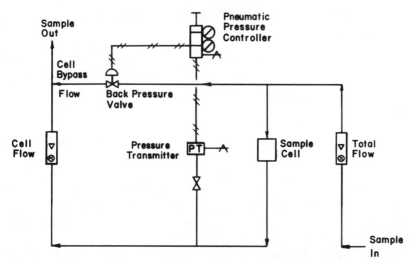

Figure 2.9. Back-pressure control scheme featuring constant cell back-pressure and constant sample flow through the cell.

also maintains a constant flow of sample through the sample cell. Upstream pressure is dropped across a throttling-type needle valve while sample cell back-pressure is automatically controlled. Back-pressure on the sample cell is set with a local pneumatic controller, and bypass flow through the back-pressure control valve is allowed to vary to maintain the setpoint. The needle valve used with the cell flow rotameter sets sample flow through the cell. Bypass flow should be at least three times greater than flow through the sample cell. The cell flow rotameter operates at the same back-pressure as the sample cell. Upstream pressure swings are compensated when the output of the controller changes, thus varying the pressure drop across the back-pressure valve.

The advantage of this sample system is its ability to control back-pressure in both the cell and bypass flow line with a minimum number of sample system components. However, if the sample stream is not reasonably dry and clean, additional preconditioning components may be required.

Process block valves and sample isolation valves have not been shown in Figure 2.9; therefore, the sample system must be shut down for sample cell cleaning and repair. A purge and calibration manifold must also be provided if zero, purge, and span gas facilities are required. A rupture disk and relief valve arrangement must also be added if the sample system components cannot withstand maximum process pressure.

High-Pressure–High-Temperature Configuration

In recent years there has been a trend toward designing high-pressure–high-temperature sample systems. These sample systems are designed to float on the process and operate at full process pressure and temperature. Process pressure and temperature must be relatively constant and not subject to load variations or pressure disturbances so that these variables will not affect analyzer calibration.

The advantage of this system is that pressure control components and relief systems are not required. There are fewer sample system components and thus fewer possible sources of sample system leaks. The disadvantage of this system is that the analyzer detector must be designed for high-pressure–high-temperature samples that are brought into the analyzer house where a sample system leak could release a tremendous amount of sample in a relatively short time.

Figure 2.10 shows a high-pressure–high-temperature sample system. Operation at pressures of up to 400 psi and temperatures of up to 300°C are possible as long as proper consideration is given to the selection of heat tracing, valves, sample cell windows, and gaskets. Electrical tracing rated at 30 W/ft or more may be required for systems operating at 300°C. Valves with metal-to-metal seats must be used because soft-seated valves will not withstand the high temperatures. Sapphire windows and graphite

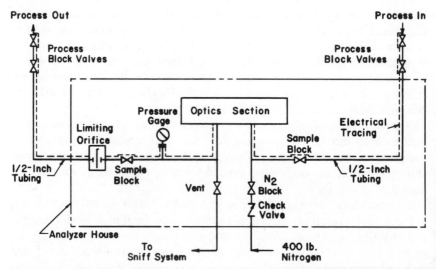

Figure 2.10. High-pressure–high-temperature sample system, featuring fewer major sample system components.

gaskets must be used in optical analyzers to tolerate the high temperature and pressure. Double process block valves are used on the process in and out lines to help assure tight shutoff even if one of the valves should leak through. Sample block valves are used to isolate the optics section, and a nitrogen purge line and vent line are located near the sample cell so that it can be quickly vented and purged.

The pressure gauge is the prime troubleshooting device. It can be used to verify that the flow-limiting orifice is not plugged, that the process block valves are holding, and also to confirm that the sample cell has been vented down properly. It can also be used to check the sample cell pressure when nitrogen is valved into the cell for purging or zeroing the analyzer. The check valve in the purge supply line prevents process material from contaminating it in cases where the process pressure is greater or the nitrogen system is vented down for repairs or maintenance.

The limiting orifice restricts the flow of process material through the analyzer. This is because it is generally undesirable to vent or bypass a substantial portion of the process stream through the analyzer system. The limiting orifice is located downstream of the sample cell because the sample cell is designed and calibrated to operate at full process pressure. All significant pressure drops must be downstream of the sample cell. The high-pressure–high-temperature sample system design does violate one of the earlier design guidelines; it does permit a relatively large quantity of process material to enter the analyzer house at a potentially high flow rate. The limiting orifice is inside the analyzer house downstream of the sample cell. When this sample system design is used, gas detectors and a ventilation fan with "fan on" light and other safety measures should be considered.

Operating and maintenance technicians should never attempt to stop leaks by tightening connectors, fittings, or cell flanges while the sample system is under pressure. The analyzer house interior could be filled with process material in a relatively short time if a line or connection were to suddenly break. Additional safety precautions are covered in Chapter 3.

2.3.3. Flow Indication and Control

The passages through a sample system flow control valve are usually very small. A typical orifice seat diameter is $\frac{3}{32}$ in. and uses a long tapered plug that is never completely withdrawn from the seat. Valve C_v values of 0.00018 to 0.02 are common for gas service with linear valve trim characteristics. Larger C_v values and equal percentage trim can be used in most liquid sample systems. Figure 2.11 shows typical lift–flow curves for equal percentage, linear, and quick opening trims on a log scale.

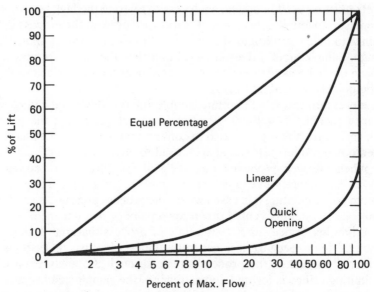

Figure 2.11. Lift–flow characteristics for equal percentage, linear, and quick-opening trims. Courtesy of Badger Meter, Research Control Valve Division.

The C_v factors are used by valve manufacturers for calculating relative flow capacities. In most sample systems a maximum gas flow of 5000 cm³/min and a maximum liquid flow of 2 gpm could be used to determine typical sample system flow valve C_v values. Equations for calculating liquid, gas, and steam flow C_v factors are as follows:

For liquids,

$$C_v = \frac{\text{gpm } \sqrt{\text{sg}}}{\sqrt{dP}} \tag{2.1}$$

For gases,

$$C_v = \frac{\text{scfh } \sqrt{\text{sg} \times (460 + t)}}{1360 \times \sqrt{P \times dP}} \tag{2.2}$$

For steam,

$$C_v = \frac{\text{pph } \sqrt{v}}{63.5\sqrt{dP}} \tag{2.3}$$

where gpm represents gallons per minute, dP is the pressure drop across valve, sg is the specific gravity of gas or liquid relative to air or water, t is Fahrenheit temperature, P is upstream pressure, scfh represents standard cubic feet per hour, and v is cubic feet per pound of steam at a specified absolute pressure and Fahrenheit temperature.

The C_v equations listed are based on standard conditions of 60°F and 14.7 psia. If air or gas temperature is other than 60°F, the correction factors shown in Table 2.5 should be used.

A pressure correction for critical flow must also be made when downstream gas or vapor pressure is less than one-half the upstream pressure. In such a situation, differential pressure dP is specified as one-half the upstream absolute pressure.

Valve manufacturers offer flow charts, nomographs, and slide rules to simplify C_v calculations. However, it is relatively easy to write a program for programmable calculators and store these equations on magnetic cards or tapes. The C_v factor for gases varies directly with the square root of the specific gravity of the gas. The specific gravities of frequently used gases are listed in Table 2.6. Figures apply at 70°F and 14.7 psia with the specific gravity of air referenced as 1.0. Specific gravities of other gases or liquids are listed in the current edition of the *CRC Handbook of Chemistry and Physics*.

Flow Indication

The most common flow control consists of a simple manual valve with throttling tip and a rotameter. The throttling valve and rotameter can be

Table 2.5. Temperature Correction

Fahrenheit Temperature	Correction Factor	Fahrenheit Temperature	Correction Factor
60	1.000	550	1.940
100	1.075	600	2.040
150	1.175	650	2.140
200	1.270	700	2.230
250	1.360	750	2.330
300	1.460	800	2.420
350	1.560	850	2.520
400	1.660	900	2.620
450	1.750	950	2.710
500	1.850	1000	2.810

Source: Courtesy of Badger Meter, Inc., Precision Products Division.

Table 2.6.   Specific Gravity of Various Gases

Name	Specific Gravity	Name	Specific Gravity
Acetylene	0.9073	Hydrogen sulfide	1.1900
Ammonia	0.5963	Krypton	2.8680
Argon	1.3796	Methane	0.5544
Boron fluoride	2.3100	Methyl chloride	1.7848
Butane (n)	2.0854	Sulfur dioxide	2.2638
Butane, iso-	2.0670	Methyl ether	1.6318
Carbon dioxide	1.5290	Natural gas	0.6000
Carbon monoxide	0.9671	Neon	0.6963
Chlorine	2.4860	Nitric oxide	1.0366
Ethane	1.0493	Nitrogen	0.9672
Ethylene	0.9749	Nitrous chloride	2.3140
Fluorine	1.3120	Nitrous oxide	1.5297
Helium	0.1380	Oxygen	1.1053
Hydrogen	0.0695	Ozone	1.6580
Hydrogen bromide	2.8189	Propane	1.5620
Hydrogen chloride	1.2678		

Source: Courtesy of Badger Meter, Inc., Precision Products Division.

combined if desired. However, this simple flow monitoring system has two significant drawbacks: (1) the rotameter is subject to plugging—when minute solids settle on the rotameter tube, they change the relative orifice area and can cause the rotameter float to stick or give false readings; and (2) pluggage may occur somewhere else in the system and may not be obvious unless the rotameter is visible.

Analyzer output can remain unchanged when the sample system pluggage occurs by trapping a static sample of process material in the lines. Extreme care must be exercised in cases where the normal output of the instrument is zero or near zero. For example, a plugged analyzer system could produce a false sense of security in someone attempting to monitor the lower explosive limit (LEL) of a potentially explosive gas mixture. The plugged monitor could not respond to a sudden increase in LEL if it were to occur, and everyone would be unaware of the unsafe condition that might lead to a fire or explosion.

For these reasons it may be desirable to install a flow transmitter or low flow alarm switch in the on-line analyzer system. The main reason for not installing flow indicators is that they are costly because additional cable from a remote field location to the receiver may be needed. Spare cables should always be pulled to analyzer locations when the equipment is originally installed. A good rule of thumb is to pull one extra cable for every 10 that are to be used.

If normal operating conditions provide on-scale analyzer output, a loss of sample flow can sometimes be noticed because the output signal may go to zero or stay where it is but appear "dead," showing no movement. Figure 2.12 shows an instrument output signal during normal and faulty operation. Most analyzer signals vary over a narrow range, and their noise characteristics soon become familiar to process operators.

The sample system flow indication should be transmitted to the control room when the analyzer is in critical service or a part of the plant's safety monitoring system. A control room technician who is alerted to the loss of sample flow will be advised that a critical analyzer system is inoperable. Analyzer maintenance technicians can then be dispatched to the field to correct the situation before an unsafe operating condition develops. Several large instrument manufacturers now market integral orifice flow assemblies suitable for use with small diameter piping and tubing. The integral orifice assembly can be used with a standard three-way manifold and flow transmitter.

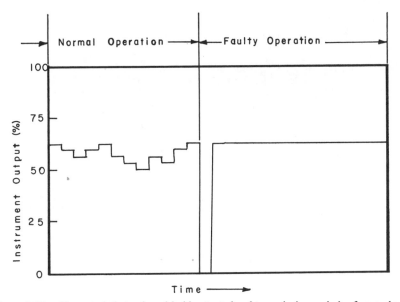

Figure 2.12. Characteristic track-and-hold output signal trace during periods of normal and abnormal operation.

Automatic Flow Control

In some cases it is desirable to provide flow control of one or more sample streams. Figure 2.13 shows a pneumatic flow control scheme for the independent control of two streams, *A* and *B*, where stream *A* is the stream of interest. Stream *B* can be used for precision dilution, catalyst addition, or the introduction of some stabilizing compound. Each flow valve and transmitter should be sized or calibrated for the expected flow range with consideration given to overrange situations. The main advantage of the independent flow transmitters is that they hold flow constant regardless of upstream pressure variations.

Many variations of Figure 2.13 are possible; it may be feasible to mix two sample streams with a simple rotameter arrangement if the upstream pressures are constant. However, a precision calibration of the rotameters will still be required. Instead of independent flow control, the designer may want to provide a ratio controller where flow $B = KA$ where K is a constant. Adequate mixing of low flows may be obtained in the tee connection, but a volumetric chamber or gas diffusion column might be required for larger streams. The rotameter shown measures the mixed gas stream or total gas flow to the analyzer cell or sensor. A pneumatic back-pressure regulator has been used to maintain a constant pressure on the sensor.

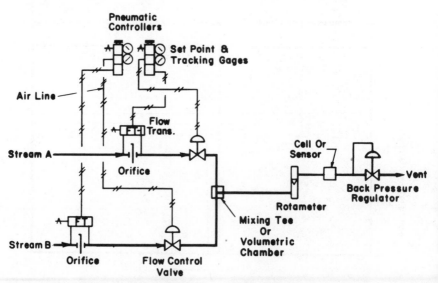

Figure 2.13. Independent flow control of two sample streams with back-pressure control after sample mixing.

The flow measuring orifice, mixing tee, rotameter, and sensor all represent potential pressure drops between the flow control valve and the back-pressure regulator. If an adequate amount of pressure drop is not provided, the independent flow controllers could interact with the back-pressure regulator.

Figure 2.14 shows a situation where a liquid sample must flow through the analyzer cell with only a few feet of equivalent water head as a motive force. Flow would be too small to measure with a conventional flow metering arrangement, so a high-velocity water jet and aspirator are used to create a vacuum to pull the sample through the system. Sample flow is controlled by the water regulator, and the combined sample and water flow is measured by the flow transmitter. The total flow signal is monitored with low flows alarmed in the central control room (CCR) because the success of this application depends on adequate aspiration. The water and sample mixture can be returned to the process at lower pressure if it is safe to do so, or it can be dumped into a waste water return system for further treatment.

Figure 2.15 shows a low-pressure vapor sampling system utilizing air aspiration. Saturated sample gas enters the top of the cooler, and con-

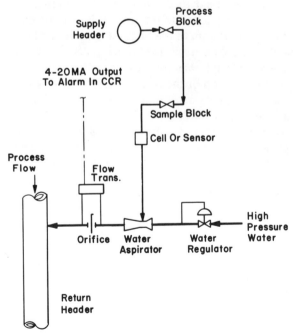

Figure 2.14. Low-pressure liquid sample system with water aspirator drive.

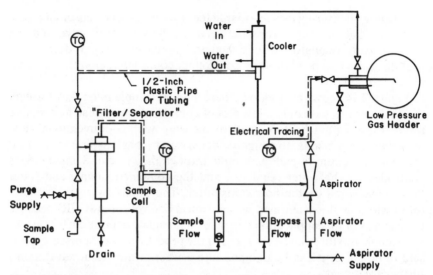

Figure 2.15. Low-pressure gas sample system with gas aspirator drive and three independent temperature control systems.

densed liquids are returned to the header. Desaturated vapors exit the bottom of the cooler and flow into the side of the filter–separator. Most of the sample leaves the bottom of the filter–separator and flows through the bypass rotameter to the air aspirator, which pulls the sample through the system and carries it back into the header. About one-tenth of the total sample flow leaves the top of the filter–separator and passes through the sample cell and sample rotameter to the air aspirator. The filter–separator coalesces mist and traps solids.

The air supply to the filter–separator is used to either purge the sample inlet line or purge the sample outlet and bypass lines. The filter–separator contains loosely packed, spun glass wool in its bottom section and contains a porous Teflon® filter in the top.

The sample system shown in Figure 2.15 uses three separate temperature controllers for controlling the temperature of the sample inlet line, sample cell, and sample outlet line. Each temperature control system must be capable of holding the sample gas temperature above its dew point.

Sample flow is adjusted by the rotameters in the system, which hold their respective lows reasonably well until air leaks or filter pluggage occurs. Low-pressure gas and liquid sample systems are the most difficult to maintain flow rates in because there seldom is enough sample system driving force or pressure for conventional flow or pressure control systems.

2.3.4. Heat Tracing and Insulation

Steam or electric tracing is often used throughout the sample system to provide freeze protection, prevent condensation of water or sample stream components, or maintain the sample at a specific detector calibration temperature. Saturated stream or low-wattage electrical tracing can be used for temperatures of up to 100°C. Temperatures above 100°C require high-pressure steam or high-wattage electrical tracing. The use of steam tracing is relatively safe in most plant atmospheres because steam does not contribute to the initiation or sustenance of fire. One disadvantage, however, is that maintenance technicians can be badly burned if they inadvertently cut into steam tracing; also, steam leaks can combine with small sample system leaks to form corrosive conditions around sample system components handling very reactive chemicals. Moisture can combine with chlorine, hydrogen chloride, or hydrogen fluoride to form extremely corrosive acids.

Electrical tracing appears to be the best choice for routine freeze protection because of the availability of low-cost self-limiting tracing. A suitable steam supply may not be readily available, or the cost of installing the tubing and traps with a condensate return system may be prohibitive. The use of electrical tracing will not eliminate the corrosion problems that can occur in sample systems handling corrosive samples since a process leak can combine with moisture in the air to form highly corrosive films. The corrosive film formed can short-circuit electrical circuits and possibly cause electrical fires.

A distinct disadvantage of electrical tracing is the shock hazard that exists if maintenance technicians accidently cut into electric tracing lines. Insulated process lines that are electrically traced should be clearly identified by "Electric tracing" signs.

Care should be given to the arrangement and routing of sample lines during the initial design phases to make sure that they are sloped to prevent trapping liquids. Extremely long sample runs should be avoided if possible. Drain taps and valves should be installed at low points in the sample system. The use of coalescing filters should be considered in wet sample streams.

Figure 2.16 shows an insulated preconditioner box with a steam coil and heat radiation plate mounted on the rear wall of the box. A steam valve with a filled system temperature controller can be used for automatic temperature control. It is not necessary to heat each sample system component inside the box when the entire preconditioner is heated. A bimetallic dial thermometer can be mounted in the door to monitor cabinet temperature. Steam or electric traced bundles can be used to heat the sample line from the process tap to the preconditioner box and from the

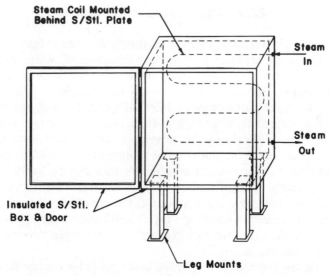

Figure 2.16.   Insulated and steam-heated sample preconditioner box.

preconditioner to the analyzer, but each application must be individually considered to determine whether steam or electric tracing is best.

Cost is an important factor in determining whether prefabricated bundles should be used or tracing and pipe insulation installed. Heat-transfer cement should be used to facilitate coupling of the heat from the tracers into the sample system line. There should be no bare sections of line showing regardless of the type of tracing and insulation installed.

For large projects, it may be more economical to buy and install custom precut electric tracing. However, exact line lengths and routing must be provided to the heat tracing supplier to ensure that all tracers match the installed lines.

2.4.   PROCESS RESPONSE AND SAMPLING TECHNIQUES

The response time of the process is the time it takes the process stream to travel from its point of origin to the analyzer tap. The point of origin could be a reaction vessel or some other point in the process where a change occurs that is to be monitored. The response time of the process could be much shorter or much longer than the residence time of the sample system. The response time of the process could be in seconds for high-velocity gas streams or in hours when long diffusion paths are involved.

Sample residence time is the time it takes for the sample to travel from the process tap to the analyzer detector. It is desirable to keep residence time to 1 min or less in most cases. Residence time of the sample depends on sample line length, filter volume, internal volume of other sample system components, and sample flow rate. Table 2.7 gives residence time for various size sample lines with a constant sample flow rate of 2000 cm³/ min at 15 psi. Table 2.7 illustrates the advantage of small-diameter tubing in reducing sample residence time and shows that the residence time for 0.75-in.-diameter tubing is comparable to thicker-walled 0.50-in.-diameter pipe. Tubing diameters are outside diameters; pipe diameters are nominal inside diameters.

Sampling Techniques

Process analyzers are used to monitor a specific point in the process for one or more specific components of interest in the process stream. When the process analyzer system is not working, laboratory samples are usually taken to monitor the process until the analyzer can be put back into service. Process variables may be automatically or manually adjusted

Table 2.7. Sample Line Residence Time[a]

Tubing Size	Residence Time
0.125 × 0.030 wall	3.9 s
0.125 × 0.020 wall	6.6 s
0.250 × 0.035 wall	33.0 s
0.375 × 0.035 wall	1.4 min
0.500 × 0.035 wall	2.9 min
0.625 × 0.049 wall	4.3 min
0.750 × 0.049 wall	6.6 min

Pipe Size	Residence Time
0.125 Schedule 40	1.1 min
0.250 Schedule 40	2.1 min
0.375 Schedule 40	3.8 min
0.500 Schedule 40	6.0 min

Source: Courtesy of Crawford Fitting Company.
[a] Per 100 ft.

based on the on-line analyzer or laboratory results. Laboratory samples are also taken when the process analyzer system is being tested or evaluated. A minimum of 30 laboratory samples can be taken to determine the linear regression line and standard deviation of x (laboratory data) and y (analyzer readings) along with the correlation coefficient of the data.

A linear regression is used to find a general mathematical relationship between two sets of variables such as laboratory analyses and analyzer readings so that the value of one variable can be used to predict the other. For example, the regression equation can be the calibration curve of the analyzer and be represented as:

$$y = mx + b \tag{2.4}$$

where x is the independent variable (laboratory data), y is the dependent variable (analyzer reading), m is the slope of the regression line, and b is the y intercept.

A term called the *correlation coefficient* is used to determine the strength of the relationship between the x and y variables and is given as:

$$r = \frac{mS_x}{S_y} \tag{2.5}$$

where m is the slope of the calibration curve, S_x is the standard deviation of the x array of data to the mean value of x, and S_y is the standard deviation of the y array of data to the mean value of y.

The correlation coefficient can range from -1 to $+1$. A correlation coefficient of 1.0 indicates a perfect correlation with the y data directly proportional to the x data. A correlation coefficient of -1 indicates that the y data are inversely proportional to the x data. Correlation coefficients of less than 0.6 are considered marginal. Correlation coefficients near zero indicate random or widely scattered data.

The standard deviation terms S_x and S_y are related to the spread of data points about the regression line. Sixty-three percent of the data points will fall within a 1 standard deviation [sometimes called 1-sigma (σ)] band when there are at least 30 data points and they follow a normal gaussian distribution curve. Thirty data points are considered a significant statistical number. Ninety-five percent of the data points will fall within a 2-sigma band about the regression line, and almost all the data points (99.75%) will fall within a 3-sigma band. For example, there is in fact more scatter in the laboratory data than the analyzer data when S_x is larger than S_y; this could indicate the need for improving sample taking or laboratory analysis techniques. A book on statistics should be consulted for a more thorough treatment of the subject. Information on linear regression and correlation coefficients is included in the instruction manuals of most manufacturers of programmable calculators.

Samples should not be taken within the confines of the analyzer house or sample system enclosure even though it is desirable to obtain a sample near the analyzer location. Additional sample taps at these locations will add valves and tubing connections that could leak and damage the sample system and analyzer electronics. However, laboratory samples must be taken at locations where those samples are the same as the sample flowing to the analyzer. Sample system taps are often located just outside the analyzer house on the line between the preconditioner and the analyzer. Sample booths may be provided at key locations throughout the manufacturing plant. Figure 2.17 shows a fiberglass sample booth with a hinged door and latch. The sample inlet line and door do not have to be sealed or gasketed when a slight negative pressure is maintained on the system by the evacuation header. The fiberglass shelf makes it easy to handle sample

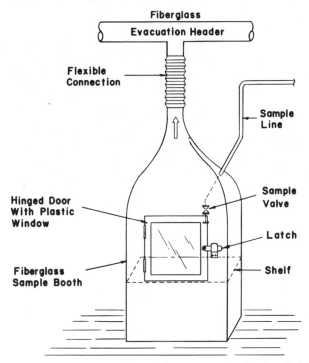

Figure 2.17. Sample booth with flexible coupling to evacuation header. Unit features see-through plastic window and work–storage shelf.

containers and provides a vented work space. The sample line entering the sample booth has a shutoff valve, tubing stub, and end cap. The end cap assures complete blockage of the sample when the booth is not in use.

The operating technician does not enter the sample booth but merely takes a sample through the sample booth door opening. Gloves, safety glasses with side shields, goggles, or a face shield may be required when taking the sample. The following precautions should be observed when taking samples and using sample cylinders:

1. Sample cylinders should be dedicated to a specific sample to prevent cross-contamination.

2. Sample cylinder valve materials should be selected so that the sample does not react with them even over long periods of time.

3. Sample cylinders with a single valve arrangement should be carefully evacuated and used shortly after evacuation to prevent dilution of the sample by air injestion.

4. The sample should be purged through the cylinder to assure that a representative sample has been obtained when the valves are closed.

In some cases special UV or IR cells with valves may be used as sample collectors. This is done to minimize the number of times that the sample has to be transferred from sample containers to laboratory analyzer detectors. Laboratory samples tend to be of equal or less concentration than the actual process material because any error in the sampling technique usually has a weakening or diluting effect on the sample. However, an opposite effect is sometimes noted, such as when high-purity water is allowed to bubble into a sample container and thus change its pH because of CO_2 absorption and carbonic acid formation.

One advantage of on-line analyzer systems is that their cells continuously sample the process stream and they cannot be affected by field sampling and handling techniques.

2.5. SAMPLE SYSTEM METALLURGY AND SYSTEM CONTAMINATION

Almost every phase of analyzer sample system design involves consideration of materials of construction. What are the materials of construction of the process pipes and vessels? Will the same materials of construction be suitable for the sample system? What is the corrosion allowance that has been designed into the plant piping system? Is this corrosion allowance suitable for thin-walled sample system tubing? Will the sample system be subjected to fume releases in the area? What affect will fume releases, humid salt air, or acid mist have on the sample system tubing? Is there a possibility of the sample system corroding from the outside in rather than the inside out?

Should preconditioner boxes be made of fiberglass-reinforced plastic (FRP) or stainless steel, or will galvanized or painted steel boxes be adequate? Is the preconditioner box hardware as durable as the preconditioner box itself? Can small-volume, porous metal filters be used with the process material, or are larger-volume cartridge filters a better choice? Will the filter material do its job without deteriorating in a relatively short period of time? Is the filter holder as durable as the filtering media? What are the materials of construction of the pressure let-down valve? How will the stem packing material hold up in this service? What affects will leaks in the preconditioner box have on the valve actuator parts? Can protective coatings be applied to any sample system components that might be subject to excessive corrosion?

What type of valve should be used in the sample system? Can soft-seated tight shutoff valves be used, or are metal to metal seats required? If soft-seated valves can be used, what elastomeric materials are acceptable? Will elastomeric materials hold up under the normal sample system pressure and temperature requirements? If an optical cell is used, what gasket material should be used to seal the windows? If Teflon® (TFE) is used, is it captive on its sides to prevent cold flow of the material? What materials are suitable for use as optical cell windows? What are the transmission characteristics of the windows, and how will they hold up to the pressure, temperature, and moisture requirements of the sample system?

Since the rupture disk is a process wetted part, is it as durable as the other sample system components? The rupture disk should have a very low (preferably negligible) corrosion rate because it is a very thin part. What are the materials of construction of the check valves? Are poppets and springs both resistant to process chemicals? Can glass tube rotameters be used? If fluorides are present, what materials will hold up in this service?

The list of questions is never-ending. While proceeding step by step through the sample system design, the designer should ask dozens of questions with regard to the selection of materials of construction. Compromises will have to be made along the way; useful life versus cost will always be an important consideration. What life expectancy is acceptable for sample system components? Can routine replacements be scheduled with the maintenance department?

Table 2.8 summarizes the chemical resistance of thermoplastic piping

Table 2.8. Chemical Resistance of Thermoplastic Piping[a]

Chemicals	PVC, Type I 73–140°F	CPVC, Type IV 73–210°F	Polyethylene 73–120°F	Polypropylene 73–180°F	Kel-F[b] 73–195°F	Teflon[c] 73–195°F	Viton[c] 73–185°F	Buna-N 73–185°F
Acetic acid (50%)	x	o	o	o	x	x	x	x
Aluminum sulfate	x	x		x	x	x	o	x
Benzene, benzol	o	o	o	o	x	x	x	o
Calcium hypochlorite	x	x	x	o	x	x	x	o
Carbon tetrachloride	o	—	o	o	x	x	x	o
Chlorine (dry gas)	o	—	x	o	x	x	x	o
Chlorine, liquid	o	o	x	—	x	x	x	o
Ethyl chloride	o	—		o	x	x	x	x
Ethylene glycol	x	x	o	o	x	x	x	x
Ferric chloride	x	x	x	x	x	x	x	x
Formic acid	o	o	x	—	x	x	x	o
Gasoline, leaded	x	—		o	x	x	x	x
Hexane	—	—		o			x	x
Hydrochloric acid (38%)	x	x	x	o	x	x	x	o
Hydrogen sulfide, dry	x	x		—	x	x	o	o

Material	1	2	3	4	5	6	7	8
Kerosene	x	x	x	x		x	x	x
Methane	x	—	—	x		x	x	x
Nitric acid (70%)	x	o	o	—		x	x	o
Phosphoric acid (50%)	x	x	—	x		x	x	o
Potassium hydroxide	x	x	x		x	x	o	o
Propane	x	—	—		x	x	x	x
Sodium carbonate	x	x	x	x	x	x	x	x
Sodium chloride	x	x	x	x	x	x	x	x
Sodium hydroxide (50%)	x	x	x	x	x	x	o	o
Sour crude oil	x	—	—				o	o
Sulfuric acid (93%)	o	o	o	o	x	x	x	o
Toluene, toluol	o	o	—	o	x	x	x	o
Trichloroethylene	o	o	o	o	o	x	x	o
Water, salt/fresh	x	x	x	x	x	x	x	x
Zinc sulfate	x	x	x	x	x	x	x	x

[a] Materials: x, recommended; —, limited service; o, unsatisfactory: blank, no data (same symbols used in Table 2.9).
[b] Trademark of Minnesota Mining & Manufacturing Company.
[c] Trademark of E. I. du Pont de Nemours & Company.

Table 2.9. Chemical Resistance of Metals

Chemicals	Carbon Steel	316 Stainless Steel	Monel	Nickel	Inconel 600	Hastelloy C	Tantalum	Titanium
Benzene	—	x	—	—	x	—	x	x
Carbon tetrachloride	—	—	—	—	x	x	x	o
Chlorine, dry gas	x	x	x	x	x	x	x	—
Chlorine, wet	o	o	o	—	o	x	x	x
Ferric chloride	o	o	o	o	o	—	x	
Fuel oil	x	x	x	x	—	x	x	
Gasoline, refined	x	x	—	—	—	x	x	
Hydrochloric acid, concentrated	o	o	o	o	o	—	x	o
Hydrochloric acid, dilute	o	o	—	—	—	—	x	—
Hydrochloric acid, dry gas	x	x	—	—	—	x	x	
Hydrogen peroxide, concentrated	o	o	—	—		x	x	x
Hydrogen peroxide, dilute	o	—	—	—	—	x	x	—
Hydrogen sulfide, dry	x	x	—	x	—	x	x	

44

Hydrogen sulfide, wet	—	x	—	x	o	x	—		—
Kerosene	x	x	x	x	x	x	x	x	
Methanol, concentrated							x	x	x
Methanol, dilute	—		—				x	x	x
Nickel chloride	o	x			x		o	o	x
Nitric acid, dilute	o		o	o	o	o	o	x	x
Nitric acid, concentrated	o			o	o			o	x
Sodium hydroxide, concentrated	—		x	x	x	x			o
Sodium hydroxide, dilute	x								x
Sodium hypochlorite	o		o	o	o	o		o	x
Sulfur, molten						x	x	x	x
Sulfur dioxide, dry	x	x	o	o	o	o	x	x	x
Sulfur dioxide, wet	o	x	o	o	o	o	x	x	
Sulfuric acid, hot concentrated	o	o	o	o			o		o
Sulfuric acid, cold concentrated	—	x	—	x			x		—
Sulfuric acid, dilute	o	o	o	o	—		x		—

Source: Adapted from Fisher *Control Valve Handbook*, copyright 1965–1977.

materials. This table should only be used only as a guide because impurities in the listed chemicals can radically change their chemical activity. These tables do not take into account stress on the piping or temperature variations greater than those shown. A corrosion engineer's handbook can be consulted for additional detailed information on the chemical resistance of plastic piping materials.

Table 2.9 shows the chemical resistance of some metals used for process equipment and piping in the chemical and petrochemical industries. These chemicals are considered at room temperature unless otherwise stated. This table does not take into account the effect of impurities in the chemicals or the affects of heat and stress on the piping. A materials engineer should be consulted in cases where limited service has been noted.

System Contamination

When a process analyzer system is about to be put into service there are a number of ways of cleaning and leak-checking the system. Water can be flushed through the system first to remove any water-soluble contaminants and gross leaks corrected at this time. A water-soluble solvent can be used after the system is drained to absorb water and remove oily deposits. Finally, hot dry nitrogen can be purged through the system to remove remaining liquids and provide an inert gas for leak detection. Final leak checking can be done from the outside with a liquid soap solution that forms large quantities of bubbles in the presence of escaping gas. All leaks in the sample system must be corrected before establishing sample flow in the analyzer sample system.

The same safety rules should be applied when working on the sample system as are applied when performing work on the process piping. Gloves and face shield or full acid suit may be required on systems that have been in operation.

Sample systems such as the one shown in Figure 2.6 require that careful attention be paid to the pressure gauge upstream of the filter that is in service when work is done in the preconditioner box. The process valve at the sample tap should be closed first when working on the process analyzer system. If the pressure on PG1 or PG2 does not drop during the venting operation, there is probably blockage downstream of the gauge. Breaking into the sample system requires great care because pockets of gas or liquid could be trapped and present a hazard. All valves ahead of the filter should be closed in case the sample system block valve leaks through, after which the vacuum valve downstream of the filter should be opened. At this time the pressure on PG1 or PG2 should be observed to drop even if the filter is almost plugged solidly. After filter pressure has been relieved, the vacuum valve can be shut and purge nitrogen valved into the system to help sweep out any remaining pockets of liquid or gas. When the nitrogen is valved in, its pressure should be observed on the filter inlet pressure gauge. Dry nitrogen should be purged through the system for 10 to 15 min before the system is vented again. After all pressure gauges confirm that the sample system pressure has been completely vented, work can proceed cautiously on the sample system parts inside the preconditioner box.

Again referring to Figure 2.6, if work is required on the sample system components inside the analyzer house, the pressure controller setpoint should be set to zero so that the control valve will close. The sample block valve just ahead of the gas manifold valves can next be closed, after which the common block and vacuum valves can be opened and the pressure on PG5 observed. The sample system pressure on PG5 should reduce to the vacuum header absolute pressure. The vacuum valve can be closed when this occurs and the sample system then purged with dry nitrogen for 10 to 15 min. After the sample outlet valve has been closed, it should be safe to work on the sample system components located between the sample block and the sample outlet valves.

CHAPTER

3

SAFETY

Safety has been given a position of prominence in this book because of its importance in process analyzer systems. It has been discussed directly and indirectly in Chapters 1 and 2 under topics dealing with analyzer accessibility, safety relief valve and rupture disk sizing; flow, pressure, and temperature ratings; and materials of construction. Likewise, it is discussed in the remainder of this book as specific analyzer system designs and applications are examined. Young engineers and chemists should use the guidelines presented here in conjunction with their company's existing safety rules and regulations.

Individual safety is a very personal thing. Most employers believe that accidents are "caused" and thus "can be prevented." In today's society many people are safer at work than at home. This is because most employers have done a reasonably good job of instilling proper safety attitudes with regard to the employee's work environment. Who has the most to lose by being unsafe? Injured individuals and their families. The employer becomes a loser, too, faced with the need to replace the injured employee with someone who must be extensively trained to perform the injured person's work. The key to a proper safety attitude is to remember that one is responsible for one's own safety.

Modern industry ranks safety equally important to productivity, quality, and profitability. If all employers felt this way, the need for regulatory agencies such as OSHA in this country and the Health and Safety Commission in Great Britain could be seriously questioned. Unfortunately, the regulatory agencies and their safeguards are still needed.

Design plays an important role in safety. Early industrial processes such as old black powder mills in this country were constructed with three huge stone walls and used a flimsy roof and fourth wall. Accidental explosions were thus directed in such a manner to prevent damage to neighboring mills and facilities. Injuries and fatalities still did occasionally occur, but losses were kept to a minimum. Although the safety of most manufacturing processes has dramatically improved in recent times, the design and layout of industrial plants, processes, and equipment is equally important today. Atomic reactors are constructed so as to contain radioactive releases if necessary.

Individual responsibility and adherence to established safety rules and regulations are mandatory. It is thus imperative that one becomes knowledgeable about the manufacturing plant and process in the immediate work area. What chemicals are being handled? What hazards are involved? Where are the safety showers, eyewash stations, stretchers, fire alarms, fire extinguishers, first aid kits, and breathing apparatus located? What safety rules apply, and where are they posted or kept? Are flame permits, burning permits, or vessel entry permits required? Must the process be valved off and vented and the electrical system locked out and tagged before commencing work? Which procedures apply for breaking into process lines, and what protective equipment must be worn? Which way is the wind blowing?

Other important safety considerations include personal protective equipment, hazardous atmosphere ratings, instrument enclosure classifications,[1] analyzer and electrical equipment design, and analyzer house design.

This chapter is not intended to take the place of the National Electric Code (NEC) or various bulletins published by the National Fire Protection Association (NFPA). Full coverage of the code is beyond the scope of this book; only a few of the more general guidelines and definitions can be covered in this work. It will always be the responsibility of the individual design engineer to provide the safest system possible to the best of personal ability.

Instrumentation using ionizing and nonionizing radiation sources for measuring density, alloy composition, or other analyses require special handling and are covered by regulations established by the U.S. Department of Health, Education, and Welfare (HEW) and various state health organizations.

Radioactive material control records and procedures are generally required for:

1. The approval and purchase of radioactive materials.
2. New installations of instruments containing radioactive sources or changes to existing instrument installations where radioactive sources are used.
3. Personnel handling radioactive sources or working in areas near radioactive sources.
4. The movement of radioactive source materials from one location to another.

5. The performance of radiographic inspection work.
6. The use of medical x-ray equipment.

A radiation protection officer (RPO) is generally appointed to supervise activities at plants where instruments using radioactive sources are installed.

3.1. OPERATING CONSIDERATIONS

From a practical standpoint it should be remembered that the sample system contains the same process material as the larger-diameter process pipes. Capacity of the system may be limited, but the same precautions should be exercised when making line breaks and venting the sample system down as those used when opening process lines. If process line breaks require the use of a face shield and rubber gloves, a sample line break in the same system should require the same personal protective equipment. If the process line break requires a full acid suit, then breaking open a sample line containing the same material should also require a full acid suit.

Since toxic, flammable, or corrosive materials are handled in some sample systems, these lines should be properly labeled identifying the chemical being transported. From a maintenance standpoint it is also helpful to show the normal direction of sample flow. All controls necessary for sampling, purging, venting, and calibrating should be readily accessible to the operations or maintenance technician. It is convenient for maintenance technicians to have access to both sides of a sample system rack or component mounting arrangement. Walkways around analyzer sample systems should have a minimum width of 24 in. without any overhead obstructions or low-level tripping hazards. In case of an emergency, a technician in the area should be able to make a rapid egress without having to worry about obstacles that could cause personal injury. Adequate clearance should also be provided around analyzers and sample systems mounted in the analyzer house.

Analyzer systems should be deenergized before attempting to make electrical repairs. In many cases a flame permit should be required if soldering irons, hot air guns, or small gas torches are needed to make repairs. Before a flame permit is issued, a survey should be made of the area to make sure that flammable vapors are not present. If work is done underground, in a pit, or in confined vessel, a portable oxygen monitor should be used to verify that it is safe for personnel to enter the area and a vessel entry permit should only then be issued. Permits of this nature generally have a limited duration, usually 8 hr or less, and must be reissued if work has not been stopped or completed in the specified period of time.

Most plants require that all sources of electrical energy be cut off, locked, and tagged before any electrical repairs are attempted. This is to prevent accidental shock and possibly fatal electrocution of anyone entering the area unexpectedly or working on the system.

The lock should be installed so that it physically prevents the switch or circuit breaker from being energized. The tags are informative in nature, and conditions authorizing their removal should be explicit. Procedures may require both the operating and maintenance groups to lock and tag the appropriate electrical disconnects. Electrical heat tracing circuits or control circuits that could possibly be energized through an external level, flow, or pressure switch should not be overlooked.

Sample systems are frequently provided with sample system taps so that grab samples may be taken at or near the analyzer location. The sample container should be clearly marked, identifying the contents, source, date, and time taken along with the technician's name or initials. Liquid samples taken in glass containers should be transported in unbreakable carriers capable of holding the entire contents of the glass container in the event of breakage. After samples have been analyzed, a safe means of disposal should be provided. The method of sample disposal will depend on whether the sample is liquid or gas, acid or base, flammable or inert, or any other special characteristic that might require consideration.

3.2. RESPONSE TO EMERGENCIES

Emergency preparedness committees are often formed to establish plant emergency procedures and set policies dealing with the training of employees who will be responding to these emergencies. These committees decide what expected or unexpected emergencies might arise and where this emergency equipment will be stored.

Emergencies may be classified as "plantwide," meaning that all personnel on the plant should be alerted and prepared to take some sort of action if necessary. The required action might be to evacuate the plant. If so, wind speed and direction would be an important factor. Other emergencies may be restricted to a specific area or manufacturing process and may be classified as an "area" emergency. In this case, personnel in the immediate area would be expected to react to the emergency.

When a fume detector or fire alarm is activated, technicians responding to the emergency should assume that the emergency condition is real until proved otherwise. Windsocks should be installed in each major production area and relative wind speed and direction noted. Persons investigating an alarm should approach the area from an upwind or crosswind direction.

Some plants require that an emergency crew respond to all emergencies. Often a special truck is used that is equipped with fire-fighting equipment and portable breathing apparatus. Each member of the crew should have specific duties and know exactly what to do in the event of a fire or fume release. Emergency crews are often trained to respond to hurricanes, tornadoes, floods, and other possible disasters as well.

In no event should a lone technician attempt to respond to a potential (area or) plant emergency. The aid of a backup person should always be enlisted, and the necessary personal protective equipment should be secured first. Depending on the emergency, fire-fighting equipment, two-way radios, portable breathing equipment, oxygen, or portable combustible gas monitors may also be required. Once the proper personal protective equipment has been obtained, the emergency team should isolate the fire or fume release and shut off the ignition source or the supply of oxygen or fuel. They should also check the area for injured or missing persons.

The manufacturing or chemical process should be so designed that isolation and an orderly plant shutdown are possible without further endangerment to personnel. Many plants also have a personnel reporting procedure so that all employees are accounted for during a plantwide emergency.

In addition to the safety aspects for fume releases, fires, and explosions, liquid spill containment and cleanup must also be kept in mind. Persons working with these chemicals should be familiar with the hazards they represent. Are they toxic, readily absorbed through the skin, corrosive, irritating, or flammable? The same precautions must be taken when cleaning up analyzer sample system spills as those taken when handling larger process spills, and the same personal protective equipment should be worn.

3.3. EQUIPMENT DESIGN

Two design concepts that seem to be increasing in popularity involve intrinsic safety and equipment purging. According to the National Electric Code (NEC). "Intrinsically safe equipment and wiring shall not be capable of releasing sufficient electrical or thermal energy under normal or adnormal conditions to cause ignition of a specific hazardous atmospheric mixture in its most easily ignited concentration.''

Manufacturers may have their individual pieces of equipment tested and certified as "intrinsically safe." This does not mean that they may be used in conjunction with the equipment of other manufacturers without additional testing and certification. The reason for this is that the entire instrument loop and its components and associated wiring must all be certified as intrinsically safe. For example, most pH probes are probably intrinsically safe and some pH monitors may be intrinsically safe, but the 115-V AC-powered recorder used to record the pH trend signals would not be intrinsically safe. With the advent of space-age electronics it has become increasingly easier to design intrinsically safe equipment.

Instrument enclosures are purged for a number of reasons. If the instrument is purged with a clean air source and has effective safeguards against purge air failure, it may be safe to use it in a hazardous area provided surface temperatures are low and the power disconnects are also suitable for hazardous locations. Many instruments in general-purpose areas are also purged to keep out corrosive fumes or moisture, thus increasing instrument life. The key to successful purging is that the purge gas source must be clean and reliable. More harm than good would be done if the purge supply were subject to become contaminated with corrosive or flammable gases.

Case grounding connections are another important consideration in instrument installations. Good grounding techniques are essential for:

1. Eliminating shock hazards.
2. Eliminating static electricity problems.
3. Minimizing RF transient signals.
4. Minimizing common mode signal potentials at instrument inputs.

Generally equipment is grounded at or near the power source. A ground rod of certain size driven to a specific depth depending on area soil conditions may be required. Structural steel is usually considered to be a good ground, but water pipes are not because of the extensive use of plastic piping in some modern water systems.

Some systems may call for a single grounding point at a centrally located computer. In this case a large grounding buss typically 500 circular mils in diameter can be grounded to the computer mainframe and looped beneath it so that a multitude of incoming signal wires may be effectively grounded at one ground potential point.

Instruments designed to analyze combustible gas mixtures may incorporate several safety features in addition to intrinsic safety. Flame arrestors designed to cool the ignited gas and extinguish the flame may be installed on the inlet and outlet sample ports. This would prevent an external flame front from entering the instrument through the sample tubing or an internal flame front from exiting the instrument. Barrier-type terminal strips are often employed to prevent electrical arcs in the event that wiring connections are made or broken while the circuits are energized.

Equipment cannot always be designed to be intrinsically safe. Likewise, sensitive electronic equipment cannot always be designed to withstand severe environmental conditions that may be present. In this circumstance analyzer systems may be placed in electrical enclosures that have been designed to withstand severe environmental conditions. Table 3.1 lists the National Electrical Manufacturer's Association (NEMA) classifications for electrical enclosures.

By definition, an explosionproof apparatus is an "Apparatus enclosed in a case that is capable or withstanding an explosion of a specific gas or vapor which may occur within it and of preventing the ignition of a specific gas or vapor surrounding the enclosure by sparks, flashes, or explosions of the gas or vapor within, and which operates at such an external temperature that a surrounding flammable atmosphere will not be ignited thereby." In England this is the same definition that is used to describe a flameproof apparatus. It should be noted that the explosionproof designation does not prevent an explosion from occurring. The case is designed to contain the explosion; the equipment inside may be totally destroyed.

Explosionproof cases usually consist of thick-walled aluminum cast-

Table 3.1. NEMA Enclosure Classification

NEMA 1	General-purpose
NEMA 2	Driptight
NEMA 3	Weatherproof
NEMA 4	Watertight
NEMA 4X	Watertight and corrosion-resistant
NEMA 5	Dusttight
NEMA 6	Submersible
NEMA 7	Hazardous (Class 1, Group A, B, C, or D)
NEMA 8	Hazardous (same as above, oil-immersed)
NEMA 9	Hazardous (Class 2, Group E, F, or G)
NEMA 10	Explosionproof (Bureau of Mines)
NEMA 11	Acid- and fume-resistant, oil-immersed
NEMA 12	Industrial

ings with gasketless covers and a multiple bolted flange arrangement. Extreme care must be exercised when using explosionproof cases to assure that the explosionproof rating is not compromised. For example, all conduits entering or leaving the case must be sealed within 18 in. of the case, and not even a single bolt may be omitted from the flanged cover! The manufacturer's installation notes and NEC requirements should be strictly followed.

A further explanation should be made concerning the terms "waterproof" and "watertight" and "dustproof" and "dusttight." A waterproof enclosure is constructed in a manner such that water will not interfere with the successful operation of the enclosed apparatus. A watertight enclosure is so constructed that water will not enter the enclosing case. Similar definitions apply to the terms "dustproof" and "dusttight." See Section 3.4 for additional information on hazardous locations.

3.4. HAZARDOUS ENVIRONMENTS

According to Article 500 of the NEC, a Class 1, Division 1 location is a location where:

(1) hazardous concentrations of flammable gases or vapors exist continuously, intermittently, or periodically under normal operating conditions; or (2) where hazardous concentrations of such gases or vapors may exist frequently because of repair or maintenance operations or because of leakage; or (3) where breakdown or faulty operation of equipment or processes might release hazardous concentrations of flammable gases or vapors and might also cause simultaneous failure of electrical equipment.

Class 1, Division 1 locations are locations where hazardous gases or vapors are handled but are normally contained in closed systems, or where hazardous concentrations are normally prevented by positive mechanical ventillation, or where electrical apparatus is located near Class 1, Division 1 locations.

Class 2 locations are locations involving combustible dust, and Class 3 locations are locations involving ignitible fibers or flyings. Divisions within these classes are similar to those described for Class 1 locations.

Equipment manufactured for use in these various classes and divisions should be clearly identified by the manufacturer. Flammable gases and vapors are grouped according to their explosive characteristics and ignition temperatures. Table 3.2 lists some of the various chemicals by these groups. Some of these gases and vapors may be toxic, flammable, or

Table 3.2.   Chemicals by Groups

Group A Atmospheres	Group D (Continued)
Acetylene	Ethane
Group B Atmospheres	Ethanol (ethyl alcohol)
Butadiene	Ethyl acetate
Ethylene oxide	Ethylene dichloride
Hydrogen	Gasoline
Manufactured gases containing	Heptanes
more than 30% v/v hydrogen	Hexanes
Propylene oxide	Isoprene
Group C Atmospheres	Methane
Acetaldehyde	Methanol
Cyclopropane	3-methyl-1-butanol
Diethyl ether	Methyl ethyl ketone
Ethylene	Methyl isobutyl ketone
Unsymmetrical dimethyl	2-Methyl-1-propanol
hydrazine	2-Methyl-2-propanol
Group D Atmospheres	Petroleum naphtha
Acetone	Octanes
Acrylonitrile	Pentanes
Ammonia	1-Pentanol
Benzene	Propane
Butane	1-Propanol
1-Butanol (butyl alcohol)	2-Propanol
2-Butanol (secondary butyl	Propylene
alcohol)	Styrene
n-Butyl acetate	Toluene
Isobutyl acetate	Vinyl acetate
	Vinyl chloride
	Xylenes

Source: Adapted from the National Electric Code.

corrosive. Refer to the gas safety data chart in Appendix 8 for further details.

Most instruments and analyzer systems used in hazardous locations are used in Class 1, Division 2, Group C or D locations. Some of the basic regulations applying to Class 1, Division 2, Group D locations are as follows:

1. Switches, circuit breakers, pushbuttons with make-and-break contacts, relays, alarm bells, and horns shall have Class 1 enclosures or be immersed in oil, hermetically sealed, or be intrinsically safe.

2. Resistor and resistance devices shall be in Class 1 enclosures or be hermetically sealed and be constructed so that exposure surface temperatures will not exceed 80% of the ignition temperature of the gas or vapor involved. Surface temperature should be clearly marked on the case of the device.

3. Fuses may be mounted in general-purpose enclosures if they do not exceed a 3-A rating at 120-V and if they are preceded by a switch in a Class 1 enclosure.

4. Electrical wires should be run in threaded rigid metal conduit or suitable armor jacketed cables. Connections to enclosures should be sealed. Conduit runs exiting to a nonhazardous location should be sealed. Sealing compounds shall not have a melting point of less that 93°C, and seal thickness shall not be less than $\frac{5}{8}$-in. Cables entering a hazardous area shall be sealed at the point of entrance.

5. Motors and generators shall be approved for Class 1 locations. In Class 1, Division 2 locations non-explosion-proof encased motors are permitted provided they are without arc-producing devices or contacts.

3.5. ANALYZER HOUSE DESIGN

Analyzer houses are used in conjunction with on-line analyzer systems for a number of reasons. If the production plant is small or the manufacturing area reasonably confined, a centrally located analyzer house can be used for all major analyzer systems in the process. This allows a good environment to be provided for both the instruments and the operations or maintenance personnel. Air conditioning, heating, and humidity control are often essential for equipment protection and stability. The analyzer house also provides a convenient means for connecting and distributing additional utility services such as power, air, water, vacuum, or calibration gas.

Large analyzer houses are sometimes constructed of brick or reinforced concrete. Sizes of up to 20 × 30 or even 50 ft are used where a large number of analyzers such as process gas chromatographs are to be housed. A recent innovation uses prefabricated steel panels with integral insulation similar to the portable steel buildings that are finding wide acceptance throughout the building industry. Analyzer shelters can be made to order depending on the total area required to house the assigned analyzers. Another significant advantage is that analyzers can be added later with much less construction work since any bulkhead fittings for tube or conduit runs need only be routed through sheet steel as opposed to drilling through a concrete wall.

Smaller-sized houses or sheds down to 6 × 8 ft in size are more frequently constructed of 2-in.-thick foam-filled walls sandwiched between reinforced plastic panels. The construction is similar to that used in modern refrigerators. The plastic materials used in smaller analyzer houses should be nonflammable and nontoxic when exposed to fire. Plastic houses can be prefabricated and are generally supplied with a bottom flange to allow them to be securely bolted to a concrete pad. When properly designed and installed, they are capable of withstanding winds of up to 140 mph.

When analyzer houses are used in hazardous locations, the equipment inside the houses should meet the appropriate electrical classification for the anticipated atmospheres. In some cases the entire analyzer house may be pressurized from a clean air source in a nonhazardous area. A slight positive pressure of about 0.1 in. of water should be maintained, and a sensitive air switch should be used to automatically disconnect all electrical power in the event of a ventilation system air failure. A good location for the sensitive air switch is in the purge air inlet supply duct to the building. Sensitive airflow switches using thermistor detectors are available that have a wide flow measuring range. The inlet air duct size and location are important because not only must an adequate turnover of purge air be delivered to the house, but its source must also come from a normally nonhazardous area. Typical designs call for 6-in. ducts to be routed 50 ft above the nearest process equipment and be connected to the suction side of the air-conditioning blower at its lowest atmospheric pressure area. This would usually be between the air filters protecting the evaporator and the blower fan assembly. Even when positive air turnover is used for the entire analyzer house, it is still good practice to use explosionproof switches and explosionproof enclosures for all power switching devices. The Instrument Society of America (ISA) has developed standard practices for installation of such purged systems.

It may be difficult to provide a clean air source or sufficient air to properly dilute flammable or toxic fumes originating from sample system leaks inside the analyzer house. When this is the case, individual pieces of equipment can be provided with purges and fume detectors can be positioned near pumps and other likely sources of leaks. If fumes are heavier than air, the fume detectors should be mounted near the floor. If fumes are lighter than air, the fume detectors should be mounted near the ceiling. Most petrochemical installations would require a combination of these mountings. When fume detector sensors are mounted near the floor, they should not be mounted where water is likely to splash on them as the result of washing down spills. Water will ruin the sensitivity of most fume detectors. This fact should also be taken into account when sensors are mounted outdoors as well. A good location for a fume detector sensor is at the entrance to the outlet air vent of the analyzer house.

Some toxic vapors that might be handled by various analyzer systems are listed in Table 3.3 along with their threshold limit values (TLVs). It should be remembered that no two people react in exactly the same way when exposed to identical quantities of these hazardous chemicals. For instance, the hazardous limit in a person who has an allergy or sensitivity to one of these chemicals may be considerably lower.

Another advantage of strategically located analyzer houses is that sample system lines can usually be kept short and additional sample components can be mounted just outside the analyzer house. The sample tap and process block valves should always be located outside the analyzer house. In some cases the analyzer house may be partitioned into two or more rooms. One room can be set up as a sample system room, another

Table 3.3. Threshold Limit Values[a] (ppm)

Chemical	Specific Gravity (Air = 1.0)	Adopted Values	
		TWA[b]	STEL[c]
Carbon dioxide	1.52	5000	15000
Carbon monoxide	0.97	50	400
Chlorine	2.45	1	3
Hydrogen chloride	1.26	5[d]	—
Hydrogen cyanide	0.94	10[d]	—
Hydrogen fluoride	0.69	3	6
Hydrogen sulfide	1.18	10	15
Sulfur dioxide	2.21	2	5

Source: Courtesy of American Conference of Governmental Industrial Hygienists, Inc.

[a] Threshold limit values are revised annually by the ACGIH; TLV booklet Publication No. 01 is available at nominal cost through AGIH Publications, 6500 Glenway Ave., Bldg. D-5, Cincinnati, Ohio 45211.

[b] The time-weighted average concentration for a normal 8-hr workday and 40-hr workweek, to which nearly all workers may be repeatedly exposed without adverse affect.

[c] The short-term exposure limit to which most workers can be exposed without harmful affect. A 15-min TWA exposure that should not be exceeded at any time during a workday.

[d] The ceiling concentration that should not be exceeded even instantaneously.

the electronics or analyzer room, and a third as a local repair facility to avoid having to move delicate or cumbersome equipment to a remote repair facility.

The sample system room should be isolated from the other rooms in the analyzer house by fume barriers and conduit seals. The sample system room is the room where process leaks are most likely to occur. There should be local or remote fume detector alarms or loss of ventilation air alarms to warn operating and maintenance technicians of unsafe conditions inside the analyzer house.

There should be two exits for each analyzer house or analyzer room. Outside exits should open outwardly at opposite ends of the house or room. All outside doors must open from the inside, and "panic bars" may be utilized to facilitate quick escapes. If air locks are employed, they should not hinder or impede escaping technicians. All doors should be equipped with eye-level clear plastic or reinforced plate glass windows. Large windows may also be used to observe certain pieces of process equipment or view possible escape routes. Some consideration should be given to the installation of break-away or kick-out panels to facilitate escapes in the event of an emergency.

The location of breathing air stations, communication systems, portable fume detectors, plant alarm repeaters, fire extinguishers, calibration gas cylinders, safety showers, and eyewash stations should also be of concern with regard to the overall design of the analyzer house facility. Figure 3.1 shows the layout of a typical analyzer house.

3.6. CALIBRATION GAS CYLINDERS

Calibration gas cylinders can present some unusual safety hazards of their own. Various mixtures of toxic or flammable gases may be used to calibrate IR analyzers, gas chromatographs, or catalytic combustion analyzers. As long as the gases are safely contained in the system, they present no hazards; however, inert gases that have similar calibration characteristics should be used for analyzer calibration whenever possible. Gas cylinders are usually under very high pressure, and if their cylinder valves are damaged, the results can be spectacular. The cylinders can become unguided missiles that can completely demolish an analyzer house or chemical laboratory.

Cylinders should be secured in an upright position in an improved cylinder rack. Restraining chains should be used to keep the cylinders in the upright position. Some benchtop holders for small gas cylinders are designed to hold the cylinder at an inclined angle. In all cases, cylinder

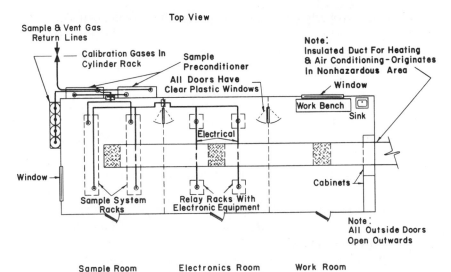

Figure 3.1. Analyzer house for multiple analyzer installation.

contents should be plainly marked on the cylinders and gas certification tags should be left attached to the cylinders. Three-part perforated service tags indicating "full," "in service," and "empty" should be placed on each cylinder and kept updated.

Compressed Gas Association (CGA) fittings should not be swapped from regulator to regulator, and the proper regulator should always be used with the specific gas cylinder. Cylinders and regulators should be treated as a single unit, and regulators should remain at their proper rack location when empty cylinders are removed. This helps to prevent contamination of regulators used on different blends of gases of a similar nature using the same CGA adapter. Consideration should also be given to the use of flame arrestors and check valves. Check valves can prevent cross-contamination of the cylinders when several cylinders are connected to the same gas header.

Additional safety precautions pertaining to gas cylinders are listed as follows:

1. Know the gas being handled. Understand its toxic, corrosive, flammable, and pressure characteristics.

2. Wear proper face and hand protection when handling or valving cylinders.

3. Keep the cylinder valve cap in position when moving or transporting cylinders.

4. Use a properly designed cylinder cart for moving cylinders.
5. Always make sure the cylinder valve is closed before replacing the protective cylinder valve cap.
6. Avoid using adapters to change fittings. Use only CGA-approved fittings.
7. Do not keep cylinders containing toxic or corrosive gases for more than 6 months.
8. Never attempt to refill a supplier's cylinder. Always return empty cylinders to the supplier.
9. Keep cylinders out of direct sunlight, and never attempt to heat a cylinder above 125°F.
10. Never tamper with the safety devices on any cylinder.
11. Remove leaking cylinders from the work area immediately and notify the supplier.
12. Always use pressure regulators to reduce cylinder pressure to safe working levels.
13. Know the location of safety showers, eyewash stations, and self-contained gas masks when working with corrosive gases.
14. Do not allow cylinders to stand in water. Elevate them with expanded steel decking if water continuously stands in the cylinder storage area.
15. Always store inert gas cylinders between cylinders of fuels or oxidizers to provide good separation between gasses such as hydrogen and oxygen, methane and air, and so on.
16. Finally remember that all gasses can cause death by asphyxiation by displacing oxygen in the lungs.

CHAPTER

4

pH AND CONDUCTIVITY MONITORS

Both pH and conductivity monitors are covered in this chapter because of their similarities. Both types of monitor are built and offered for sale by the same manufacturers. Both types are housed in similarly sized NEMA 3, 4, or 4X enclosures suitable for pipe stand installations or panel mounting. Not only do the electronic units look alike, but their insertion, submersion, and flow-through probe assemblies also are similar. Because of the common use of enclosures, plug-in circuit boards, and manufacturing techniques, the costs of the probe assemblies and electronics units are also about the same for either type of monitor. However, pH monitors with integral flow indication, three-mode controllers, or microprocessors cost several thousand dollars more.

All pH monitors are designed to measure the hydrogen ion concentration for determining the strength of various aqueous solutions of acids, bases, or salts. Here the similarity ends.

The pH measuring electrodes are high-impedance devices ranging from 60 to 400 megohms, and pH monitors are high-impedance voltmeters with input impedances ranging from 10^{12} to 10^{13} Ω. By contrast, conductivity monitors measure the specific conductance of a solution that is the reciprocal of the resistance of a 1-cm cube of solution. Specific resistance is given in Ω-cm, and conductance is given in S/cm. Solution conductivities typically range from 1 to 1 million μS/cm. A resistance bridge with a conductivity cell as one leg of the bridge, powered by an AC source and having a readout device, is the simplest form of a conductivity monitor. The conductivity probe has two electrodes rigidly set in an insulating material. The cell constant is determined by the electrode area and spacing. The cell is not affected by internal volume or the presence of nearby metal objects. The resistance bridge is a relatively low impedance device.

4.1. pH MONITORS

During the past 10 years several improvements have been made in pH monitoring and process control. Some of these improvements include:

63

1. The development of probe designs that minimize electrode fouling.
2. The development of stable, high-impedance, unity voltage gain, zero-offset current preamplifiers.
3. The development of universal probe holder designs suitable for both submersion and flow-through applications.
4. The development of gel-filled, long-life reference electrodes with ceramic junctions.
5. The development of Hastelloy C® encased ultrasonic cleaners for use in active chemical systems.
6. The development of better electrode contact cleaning devices such as rotating nylon brushes and swirling Teflon® balls.
7. The incorporation of flow monitors for feed-forward process control concepts.
8. The incorporation of microprocessors for characterization of the pH signal to match the titration curve of the process.

The safety of pH systems has also been improved so that with the proper options they can be used in most hazardous areas. Hermetically sealed relays, circuit boards potted on one side, and connectors using contacts with high contact pressure are used in addition to intrinsically safe circuits and corrosion-resistant enclosures to extend component and instrument life.

4.1.1. Theory of Operation

All acids, bases, and salts are electrolytes. In a water solution the electrolytes dissociate more or less completely into electrically charged ions.[1,2] This dissociation increases with increasing dilution and approaches a complete dissociation limit. The number of ions actually formed depends on the characteristics of the electrolyte, including its concentration and temperature. Strong electrolytes are completely ionized in water solutions, leaving no undissociated molecules behind.

In electrolysis current is carried by the ions and the conductivity of the solution is proportional to the number and type of ions in it. Water is the most frequently used solvent for industrial acids, bases, and salts. Water ionizes to form positive hydrogen ions and negative hydroxyl ions, and these ions also continuously recombine to form water. At equilibrium conditions, the speed of reaction in both directions is the same.

For pure water and dilute aqueous solutions, the ion product of the hydrogen and hydroxyl ions is constant and equal to the equilibrium constant of 10^{-14} at 25°C. This means that there are 10^{-14} moles/liter of hydrogen and hydroxyl activity in a liter of water at 25°C. In the case of pure water the concentration of hydrogen and hydroxyl ions are equal at 25°C and the water is chemically neutral, neither acidic nor basic. Therefore, there are 10^{-7} moles/liter of hydrogen ion activity and 10^{-7} moles/ liter of hydroxyl activity in a liter of pure water.

Table 4.1 illustrates that the concentration of the hydrogen and hydroxyl ions are equal only at the neutral point and that both ions are present at all other points in the table. Furthermore, the product of the two concentrations is equal to 10^{-14} under ideal conditions. Since the hydrogen ions are always present in the solution, their concentration is a precise measure of acidity or alkalinity. Because negative exponentials are difficult to work with, Sorensen developed the pH scale in 1909 using the properties of logarithms to the base 10 as follows:

$$(H^+) = 10^{-x} \tag{4.1}$$

$$X = -\log(H^+) \tag{4.2}$$

$$pH = -\log(H^+) \tag{4.3}$$

$$pH = \frac{1}{(H^+)} \tag{4.4}$$

Table 4.1. Relationship of pH to Ion Activity and Molarity

	pH	Ion Concentration, Moles/Liter	
		[H$^+$]	[OH$^-$]
Acidic	0	1.0	0.00000000000001
	1	0.1	0.0000000000001
	2	0.01	0.000000000001
	3	0.001	0.00000000001
	4	0.0001	0.0000000001
	5	0.00001	0.000000001
	6	0.000001	0.00000001
Neutral	7	0.0000001	0.0000001
Basic	8	0.00000001	0.000001
	9	0.000000001	0.00001
	10	0.0000000001	0.0001
	11	0.00000000001	0.001
	12	0.000000000001	0.01
	13	0.0000000000001	0.1
	14	0.00000000000001	1.0

As shown in equation (4.3), pH is the negative log of the hydrogen ion activity. However, while some have said that the "p" in pH has come to mean the term "−log," it actually stands for the German word "portenz," meaning "power of" 10 for x as shown in equation (4.1).

It should be remembered that the pH scale represents active acidity. The pH scale is logarithmic; a solution pH of 3 is 10 times more acid than a solution with a pH of 4 and 100 times as acid as a solution with a pH of 5. "Active acidity" refers to the hydrogen ion concentration of a solution. Weak acids of the same normality as strong acids are not as strong because they are not completely dissociated in solution. For example, 0.1 N acetic acid solution has a pH of 2.9 compared to 0.1 N hydrochloric acid with a pH of 1.0.

The (H^+) term in equations (4.1) through (4.4) represents the gram formula weight, which in this case is numerically equal to grams per liter. A 1 N solution contains 1 gram equivalent weight of replaceable hydrogen or hydroxyl groups per liter of solution.

In most process applications the exact chemical composition of the process stream may not be known. Strong bases may be added to strong acids or vice versa to neutralize the stream or adjust it to the desired pH value. Figure 4.1 shows the titration curve obtained when titrating a

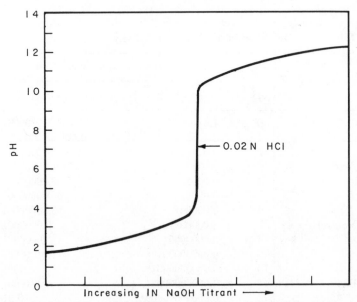

Figure 4.1. Titration curve for titrating a strong acid with a strong base. Courtesy of Fischer & Porter Company.

strong acid with a strong base. When the acid reaches the neutral point of pH 7, all the acid and base have been converted to the harmless products of neutralization; water and salt, in accordance with equation (4.5):

$$HCl + NaOH \rightarrow NaCl + H_2O \tag{4.5}$$

Figure 4.1 shows that this particular titration curve is symmetrical and difficult to control in the pH range of 4 to 10. The addition of buffers can greatly change the slope of the pH curve, thereby making control of a specific pH value easier to accomplish.[3] Buffers are mixtures of weak acids and weak acid salts or weak bases and weak base salts. Carbonates are the most useful and common buffers. Figure 4.2 shows how a buffer can change the slope of the titration curve.

For pH measurements, a glass or measuring electrode and reference or half cell make up the sensing element as shown in Figure 4.3. Their zero potential and isopotential point is at 7.0 pH. The reference electrode maintains electrical continuity with the process and establishes a reference or constant potential regardless of the wide range of pH variations that might occur in the process stream. The glass or measuring electrode

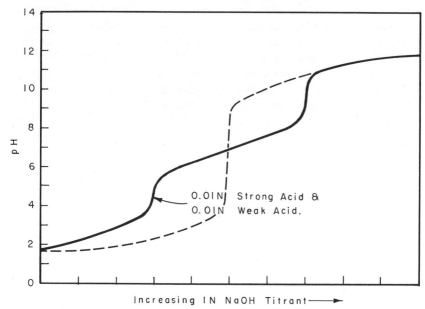

Increasing 1N NaOH Titrant ⟶

Figure 4.2. Titration curve showing the effect of buffering a strong acid with a weak acid. Courtesy of Fischer & Porter Company.

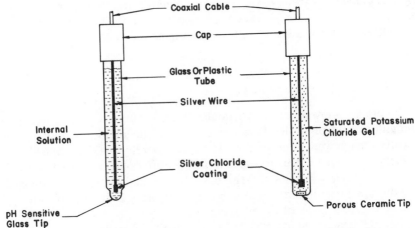

Figure 4.3. Glass pH measuring and reference electrodes. Courtesy of Uniloc/Rosemount Company.

is the part of the sensor that develops a small potential (millivoltage) proportional to the change in hydrogen ion activity. This potential is approximately 59.7 mV per pH unit at 25°C. The internal solution of the measuring electrode is captive and never needs to be replaced. However, the solution in the reference electrode will eventually be depleted and the electrode will have to be replaced or refilled in 9 months to 1 yr. Gel-filled electrodes do not expend their internal fill, thus minimizing electrode maintenance. Flowing reference electrodes are less likely to foul but are less frequently used because of their high electrolyte consumption.

Both electrodes must make good contact with the process stream for the sensor to work properly. The tips of the electrodes are usually kept wet because the actual ionic transfer takes place in a molecular layer of hydrated glass at the tip. If electrode tips are permitted to dry out, it may take several hours for the hydrated layer to reform. During this time considerable electrode drift may be encountered.

Table 4.2 shows typical sensor output in millivolts as a function of pH and temperature.

4.1.2. Calibration Techniques

There are two ways to calibrate pH monitors. However, certain electrical adjustments are usually made prior to calibration by either method. The initial electrical adjustments may include adjusting the bias of a high-impedance amplifier for zero offset current when the input is short-cir-

Table 4.2. Sensor Output as a Function of pH and Temperature[a]

	Millivolts vs. Temperature (°C)			
pH	15	25	50	80
0	396	410	444	486
1	340	351	381	416
2	283	293	317	347
3	226	234	254	277
4	170	176	190	208
5	113	117	127	139
6	57	59	64	69
7	0	0	0	0
8	−57	−59	−64	−69
9	−113	−117	−127	−139
10	−170	−176	−190	−208
11	−226	−234	−254	−277
12	−283	−293	−317	−347
13	−340	−351	−381	−416
14	−396	−410	−444	−486

Source: Courtesy of Uniloc Division, Rosemount, Inc.
[a] Typical electrode response at 99% efficiency.

cuited. Then the span or gain of the main amplifier can be adjusted so that a positive or negative 410-mV input causes a meter deflection or digital reading corresponding to pH values of 0 or 14 at 25°C. Some pH monitors are equipped with a manual temperature compensator so that any temperature between 15 and 85°C can also be simulated.

Calibration with known buffer solutions is the first, oldest, and most important method for calibrating pH monitors because it checks the electrodes as well as the cabling and electronics of the monitor. A calibration with known buffers is performed by removing the probes from the process and alternately submersing them in buffer solutions of 4 and 10. The probes should be rinsed with distilled water between buffer checks to prevent cross-contamination of the buffer solutions.

The standardization control is adjusted while the probe is immersed in the buffer so that the probe reads within 0.1 pH unit of the value of the buffer. Any two buffer solutions with a reasonable span difference can be used. The standardized control compensates for minor electrical differences encountered during probe manufacturing and probe aging. If the probe cannot be standardized to read the buffer value within the specified accuracy, the solutions are usually discarded and new probes prepared for use.

Front panel meters usually display the full 0 to 14 pH range. However, instrument output can usually be varied by zero suppression and span elevation controls on the output card or printed circuit board. Full-scale output spans corresponding to 2, 4, or 10 pH units are available depending on the manufacturer. When the full-scale output corresponds to two pH units, any two pH units such as 0 to 2, 3 to 5, or 12 to 14 can be used. With a "universal output" card, the instrument output can be 1 to 5 V, 4 to 20 mA, or 10 to 50 mA depending on the output mode selected.

Probe simulators can be used to check preamplifiers, cables, amplifiers, and output circuits. This is the second method that can be used to calibrate pH monitors. However, it should be remembered that probe simulators cannot be used to check the operability of the measuring and reference electrodes. When outputs other than 0 to 14 pH are required, the probe simulator is a valuable tool. For example, if a 4- to 20-mA output is wanted for a pH range of 10 to 12, the probe simulator would be set on 10 and the zero control of the output circuit would be adjusted to produce a 4-mA output. The probe simulator would then be set to 12 pH and the span control of the output circuit would be adjusted to produce a 20-mA output. Very fine control could then be achieved for a process whose normal pH control point would be in the 10- to 12-pH range. Figure 4.4 shows a typical probe simulator.

4.1.3. Design Features and Applications

Typical measuring electrodes are designed for process pressures of up to 100 psi and temperatures of up to 100°C. Specially constructed probes can be used at pressures of up to 150 psi and temperatures of up to 130°C.

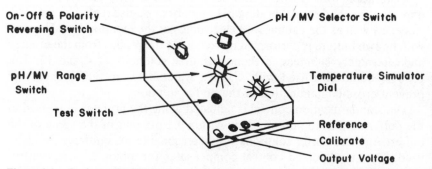

Figure 4.4. Typical pH–millivolt calibration box. Courtesy of Uniloc/Rosemount Company.

Table 4.3 shows how the pressure rating must be decreased as temperature increases. Measuring electrode resistance can be 75, 150, 300, or 400 megohms. In general, probes designed for high-temperature use have the highest probe resistance. Probes designed for use in the −5 to 40°C range have the lowest resistance. The pH probe resistance doubles for every 7°C drop in temperature below 25°C. The internal element of the measuring electrode is silver–silver chloride, and the tip of the electrode is made of specially formulated glass that is very responsive to the hydrogen ion. Probe bodies are made of Ryton®, epoxy, and Viton®.

Hydrogen ion exchange at the electrode tip generates very small currents in the order of 3×10^{-10} A per pH unit. To keep measuring errors in the 0.1 to 1.0% range, the input impedance of the preamplifier must be 10^{12} to 10^{13} Ω. The ideal preamplifier has high current gain, unity voltage gain, and low offset or bias current, typically less than 10^{-12} A at 25°C.

Modern reference electrodes also have Ryton® bodies and silver–silver chloride internal elements. They are filled with a KCl gel or slurry electrolyte. The gel minimizes electrode contamination by the process stream. Various configurations are available equipped with either replaceable or nonreplaceable wood or ceramic junctions or plugs. The wood junction gradually deteriorates when used continuously in solutions below 4 or above 10 pH. The typical resistance of a reference electrode can be anything from a few to 300 Ω, and they are generally designed to have superior resistance to clogging, stray potentials, and streaming effects. Where sulfur compounds, strong oxidizing or reducing agents, or heavy metals tend to produce reference contamination, double-junction reference electrodes may be utilized. The internal reference electrode contains KCl–AgCl, and the outer solution contains an inert salt such as potassium chloride or potassium nitrate. The double-junction reference electrode does not eliminate contamination problems, but it does significantly extend probe life.

Table 4.3. Pressure-Temperature Relationship for Polyvinyl Dichloride and Stainless Steel Probes[a]

Polyvinyl Dichloride		Stainless Steel	
Pressure, psi	Temperature, °C	Pressure, psi	Temperature, °C
100	60	100	60
50	80	75	100
30	100	50	130

[a] Other probe body materials may differ significantly.

If the reference electrode is the flowing type, its internal pressure must always be kept a few pounds per square inch above the process pressure to prevent contamination by the process. Combination probes similar to the one shown in Figure 4.5 with annular ceramic reference junctions are especially effective when making continuous measurements in the presence of high potential gradients, streaming potentials, or high solution currents such as encountered in plating baths or pH measurements made in extremely high purity water.

Flow-through-type probe holder arrangements similar to the one shown in Figure 4.6(a) are designed for maximum pipeline flow rates of 7 to 10 fps or 15 to 20 gpm. At lower flow rates there is no self-cleaning effect, and at higher flow rates electrode erosion can occur. For best cleaning action, the reference electrode should be upstream of the measuring electrode. Operation at 150 psi is typical; test pressure is 300 psi at 25°C.

The flow-powered cleaner shown in Figure 4.6(b) uses three rotating Teflon® balls to reduce electrode fouling.[5] The process sample enters the cell tangentially and a vortex is created, causing the Teflon® balls to rotate in a circular swirling motion. The Teflon® balls impinge on the electrodes helping to keep them clean. The recommended flow rate for this type of cell is 5 to 15 gpm.

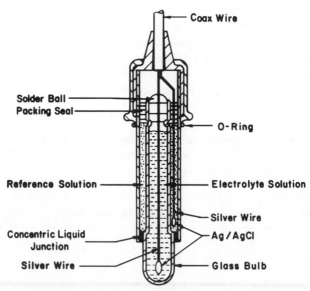

Figure 4.5. Combination pH probe. Courtesy of Uniloc/Rosemount Company.

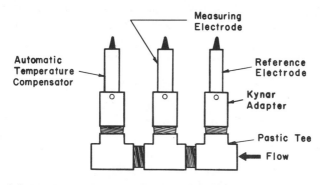

(A) Process Flow Through Cell

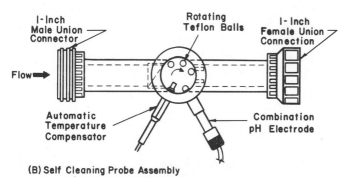

(B) Self Cleaning Probe Assembly

Figure 4.6. (a) Self-cleaning, flow-through pH probe holder featuring twist-lock-type pH probes. Available from Uniloc/Rosemount and Van London Companies. (b) Self-cleaning probe holder featuring rotating Teflon balls. Courtesy of Uniloc/Rosemount Company.

Most pH measuring systems exhibit typical electrode response of less than 1 s, excluding any problems relevant to probe fouling. One preamplifier design permits accurate signal transmission over distances in excess of 3 miles without automatic temperature compensation (ATC) and 1000 ft when equipped with ATC. The 1000-ft distance appears to be becoming standard with many manufacturers. Modern pH meter designs are also immune to ground loop noise and radio-frequency interference (RFI). Whereas long distances are possible between the preamplifier and the controller, they cannot be tolerated between the electrodes and the preamplifier. Shielded or guard shield designs must be used to prevent pickup of RFI and other stray electrical transients.

Overall system stability is excellent—typically 0.1 to 0.001 pH unit change per week. Temperature coefficients of 0.001 pH unit/°C are also typical. New microprocessor-equipped pH monitors are able to characterize the pH signal to match the titration curve of the process, thereby linearizing the curve and permitting precise control. Characterization is possible even in cases where the titration curve is not symmetrical. The microprocessor-equipped monitor can combine pH indicator, temperature and flow monitors, and a three-mode process controller in a single case. The microprocessor also provides some diagnostics such as electrode or circuit failure alarms. By providing closer control in existing processes, chemical consumption and operating costs can be reduced.

One application for pH monitors is measuring the pH of the heat-transfer medium exiting the shell side of process heat exchangers to detect process leaks into the cooling or heating medium. Figure 4.7 shows a typical pH meter pipe stand installation where the pH of a refrigerant is being monitored. Since the refrigerant is too cold for the automatic or manual temperature compensator of the pH meter, a small single-pass tubing-type heat exchanger has been installed on the sample inlet line to warm the refrigerant to about 50°C. Steam is used to heat the sample refrigerant; however, in some applications steam condensate might be

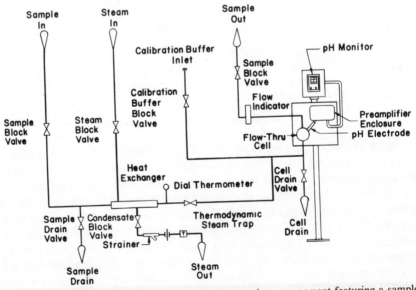

Figure 4.7. Diagram of pH monitor pipe stand mounting arrangement featuring a sample system heat exchanger and calibration inlet.

adequate. Note that provisions for adding buffer solutions are provided for pH monitor calibration.

If it was desirable to cool the sample, process or cooling tower water could be used in a standard heat-exchanger configuration. The purpose of the simple, single-pass sample heat exchanger shown is to heat or cool the sample to bring it within the ATC range of the pH meter. A temperature sensor and temperature-controlled valve would have to be added in order to have the heat exchanger actually regulate the temperature of the sample.

Figure 4.8 shows the single-pass heat-exchanger details. The unit shown is an all–stainless steel unit with the sample transported by the $\frac{1}{2}$-in. tubing and the heating or cooling medium inside the 2-in. pipe flowing in a countercurrent direction to the sample.

Another use of pH monitors is measurement of plant effluents returned to the ecosystem such as industrial waste treatment plant effluents, sewage treatment plant outfall, and waste hold pond outlets. The acceptable pH range is 6 to 9 pH for most of these applications. Some other pH applications include:

1. *Electrolysis.* Combination electrodes are preferred because of the millivolt potential that would exist between separate electrodes. Since currents are not constant, consistent readings can be made only with separate electrodes on a sample removed from the current path.

2. *Neutralization of electroplating wastes.* Again, a combination electrode would be preferred for operation in electroplating baths.

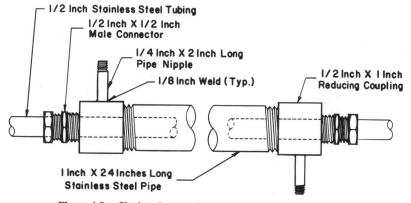

Figure 4.8.   Single-tube sample system heat-exchanger details.

3. *Cooling tower inhibitor studies.* Acidic conditions lead to corrosion, and basic conditions lead to scaling. Minimization of such extremes can be achieved through pH control of cooling tower water.
4. *Fermentation processes.* Here sterilizable electrodes made of glass or stainless steel are preferred for pharmaceutical processes involving fermentation. Periodic sterilization of the electrodes is required.
5. *Processing of chemicals, paper, beer, soap, and ores.* High-temperature phosphate baths used in capacitor manufacturing require the use of high-temperature electrodes.
6. *High-purity water studies.* These applications require electrodes insensitive to streaming potentials caused by varying flow rates of high-purity water past the electrode tips.

4.1.4. General Comments

Poisoning of flowing reference electrodes and fouling of nonflowing reference and measuring electrodes continue to be the problems most frequently encountered in industry. Improved probe designs and self-cleaning accessories have helped, but occasional probe cleaning with 4 to 10% aqueous hydrochloric acid is still necessary. Hot water and dilute acids will not damage pH probes, but other solvents may have to be used to remove fouling depending on the nature of the fouling agent.

Most manufacturers state that the response time of their system is less than 1 s. However, response time is drastically affected by static films on the measuring electrode surface.[6,7] When fouling occurs, the pH electrode measures the hydrogen ion concentration of the fouling barrier. A film of only a few millimeters can increase the response time of a measuring electrode by a factor of 40! Reference electrode fouling causes gradual drifts in calibration rather than sluggish probe response. When both probes are fouled, a sluggish, drifting system results. High-stability electronics and an adequate sample system are of little help in this case.

Another frequent cause of pH sensor failure is mechanical damage to the very thin, sensitive glass tip of the measuring electrode. Any scratches, and especially any minute cracks, are sufficient to reduce the output and sensitivity of the measuring electrode to a useless value. Very careful handling and inspection are required to avoid damage to and ensure the integrity of the sensitive tip area of the measuring electrode. Extreme care must be exercised to minimize the short-circuiting effects that even oil from fingerprints can have on the high-impedance side of the pH measuring system; and surfaces from the measuring electrode to the preamplifier input require isolation from ground in the order of thousands of megohms to ensure accuracy. Cotton swabs immersed in methanol or a similar nonresidual solvent are useful in removing such contaminants after probe installation has been completed.

Another area where considerably more research is being done is in the field of low-temperature measurements. Low-temperature, low-resistance commercial probes are effective to −5°C. The development of pH probes suitable for use to −35°C would greatly reduce the sample preconditioning requirements in refrigeration systems.

Flat-tip, all-glass pH electrodes are a recent development for the medical industry. These probes are sterilizable and can be used for skin contact measurements.

4.2. CONDUCTIVITY MONITORS

Ion current theory and the concept of solution conductivity have been known and applied for more than 80 yr. Improvements during the past 10 yr appear subtle at first, but they have been significant. Recent trends and innovations include:

1. A greatly diminished use of direct coupled, line frequency AC Wheatstone bridge circuits.
2. A wider selection of electrode and sensor body materials.
3. High-stability solid-state electronics with input and output signal isolation.
4. Failsafe alarms and NEMA enclosures suitable for use in hazardous locations.
5. The design of special instruments for special applications, including ratio monitors, rinse control systems, salinity monitors, and boiler blowdown control systems.
6. Increased use of electrodeless conductivity systems in highly conductive, highly corrosive, highly abrasive, or radioactive applications.
7. The beginning of a trend to calibrate conductivity instruments in terms of microSiemens per centimeter instead of microMhos per centimeter.

4.2.1. Theory of Operation

Electrolytic conductivity is a measure of the ability of a solution to carry an electric current. Electric current is carried by the ions in a solution that

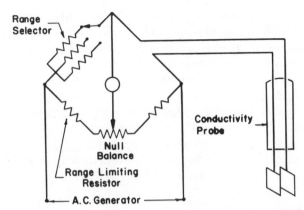

Figure 4.9. Basic conductivity probe and bridge details. Material supplied by Beckman Instruments, Inc., Cedar Grove operations. Copyright 1981, Beckman Instruments, Inc.

are produced by acids, bases, or salts. The specific conductane of a solution is the conductivity of a 1-cm cube of solution at a specified temperature, specifically 25°C at standard conditions. The specific resistance is the resistance of a 1-cm cube of solution and is given in ohms-centimeters. The specific conductane is given in microSiemens per centimeter.

Figure 4.9 shows a basic conductivity monitor consisting of a conductivity probe or sensor, AC Wheatstone bridge circuit, null detector, and power supply. Manufacturers use power supplies that provide frequencies ranging from 4 to 1000 Hz. Higher bridge frequencies are usually used for highly conductive solutions. The conductivity range of solutions typically varies over eight decades. Probes and bridges are usually selected for the range of interest, with a selectable three-decade instrument commonly employed.

Table 4.4 illustrates the range of conductivity versus resistivity measurements. Figure 4.10 shows the block diagram of a modern solid-state conductivity system. For portable intrinsically safe operation, the isolation transformer and bridge rectifiers can be replaced with an energy-limited battery pack or energy-limited barrier module.

The conductivity probe or sensor consists of two or more metal plates, rings, or rods firmly positioned within an insulated chamber. The chamber isolates the sample from nearby metal parts such as pipes and tanks. Polarizing effects are minimized by using materials not subject to polarization or by using a coating of spongy black platinum on the electrode surfaces. Modern conductivity probes have thermistor- or temperature-sensitive resistor networks built into them for automatic temperature compensation.

Table 4.4. Typical Conductivity Values

	Conductivity, μS/cm	Resistivity, ohm-cm
Distilled water	1	1M
Raw water	50	20K
Seawater	20K[a]	50
30% H_2SO_4	1M[b]	1
Contacting electrode range	0.05 to 20K	20M to 100
Electrodeless system range	1K to 1M	1K to 1

Source: Material supplied by Beckman Instruments, Inc., Cedar Grove Operations. Copyright 1981, Beckman Instruments, Inc.
[a] K = 1000.
[b] M = 1 × 10^6.

For highly conductive solutions, such as those that are particularly susceptible to electrochemical polarization, electrodeless conductivity systems are recommended.[8] The electrodeless system now offered by several manufacturers is particularly advantageous in processes where it is difficult, if not impossible, to use electrodes in direct contact with process solutions. Two examples are where abrasive solutions are processed and where radioactive material recovery operations can become contaminated with radioactive materials.[9]

As shown if Figure 4.11, the electrodeless conductivity system measures the resistance of a closed loop of solution coupling two toroidially wound coils that have been encapsulated in close proximity. An oscillator produces an AC drive signal in one toroid, thus generating a current in the solution loop. The solution current induces a current in the second toroid, and this signal is rectified or otherwise detected, amplified, and displayed on a receiver. Isolated output currents or voltages can be provided. The effective cell constant of the toroidal coil arrangement is essentially the ratio of the length of the cell to its cross-sectional area. Flow-through and dip-type or immersion-type sensors are available.

Specialized conductivity monitors for low-conductivity applications have become known as *water-purity* or *water-quality monitors* where any form of ion contamination is considered an impurity. A recent development to increase the sensitivity of this kind of contaminant or impurity detector is the inclusion of an easily replaceable ion-exchange cartridge in the sample stream immediately upstream of the conductivity cell. Very

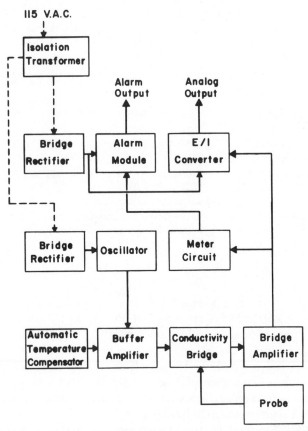

Figure 4.10. Block diagram of a modern conductivity analyzer. Courtesy of Milton Roy Company.

minor leaks of fluids such as treated cooling tower water into high-purity boiler condensate samples might eventually become diluted to the point that even a conductivity monitor with a cell constant of 1.0 or 0.1 would not sense. The ion-exchange cartridge, however, exchanges sodium ions for each mineral ion adhering to the resin in the cartridge. The conductivity of the cartridge effluent is in effect amplified, and the much larger number of sodium ions released tends to allow the conductivity monitor to respond more than it would without the ion-exchange cartridge. Conductivity alarm setpoints of 0.5 or 1.0 μS enable the quick detection of leaking heat exchangers before serious damage can occur in systems dependent on the return of high-purity steam condensate or boiler feedwater. Another specialized form of conductivity monitor is the total

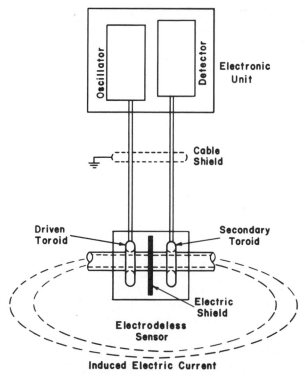

Figure 4.11. The electrodeless conductivity system featuring noncontacting electrodes. Courtesy of Foxboro Company.

dissolved solids analyzer. Table 4.5 shows the relationship between resistivity, conductivity, and dissolved solids.

4.2.2. Calibration Techniques

There are two common calibration techniques. Conductivity standards, having a known conductivity and accuracy, can be purchased from suppliers of reagent-grade chemicals. Standards can be supplied in convenient 1-liter containers, but 1-gal or 5-gal sizes may be more economical. A shelf life of 6 months is typical. Conductivity solutions can be similar to pH buffer solutions in composition, but both the weak acids or bases and salts will contribute to the total solution conductivity.

Since conductivity monitors actually measure solution resistivity with an AC Wheatstone bridge, cell simulators can be made with precision resistor sets. Usually the resistors are mounted in an electrical connector

Table 4.5.   Resistivity and Conductivity versus Dissolved Solids

Conductivity, μS/cm	Resistivity, ohm-cm	Dissolved Solids	
		ppm Salt	ppm Carbonate
0.1	10M	0.02	0.015
1.0	1M	0.42	0.30
10	100K	4.8	4.0
100	10K	48	40
1000	1K	480	400
10K	100	4.8K	4.0K
100K	10	—	—

[a] K = 1000.
[b] M = 1 × 10^6.

that will connect to the electronics unit or take the place of the conductivity sensor so that the entire electronic unit and probe cable can be tested. Resistors are usually matched to within 1% and their value chosen to produce a midscale meter reading on the range of interest.

Conductivity standards are available to check the response of the conductivity probe, the cable, and the calibration of the electronics unit. The cell simulator checks only the operability of the cable and electronics unit.

4.2.3.   Design Features and Applications

Conductivity monitors are available in NEMA 3, 4, and 12 enclosures for hazardous atmospheres. Modern units feature solid-state circuitry that is feedback stabilized and immune to RFI. Some AC bridges are powered by a square-wave generator to improve linearity and accuracy and to increase rangeability. Plug-in options and failsafe alarms are standard with many manufacturers. Input and output signal isolation are usually standard features of most modern conductivity monitors.

Alternating-current Wheatstone bridges are powered by 60-, 75-, 85-, 250-, 320-, 800-, or 1000-Hz oscillators depending on the manufacturer. A wide variety of voltage and current outputs are available. Voltage outputs include 0 to 10 mV, 0 to 100 mV, 0 to 1 V, or 0 to 10 V. Current outputs of 0 to 1, 0 to 4, 1 to 5, 0 to 16, 0 to 20, 4 to 20, 5 to 25, 0 to 40, (or) 10 to 50 mA are available. The best way to purchase a conductivity monitor with the output range required is to specify that range and consult the vendor for the appropriate output circuit configuration. There is a trend toward using an isolated 4- to 20-mA output signal as an Instrument Society of America (ISA) standard. This output provides a "live zero," which is useful when troubleshooting defective equipment. A further lack of uniformity is also evident in the range of ATCs available. Again, depending on manufacturer, temperature compensation ranges of 0 to 100, 5 to 50, 50 to 125, 100 to 200, or 5 to 85°C are available. The need for automatic temperature compensation is demonstrated in Table 4.6.

Table 4.6 also shows that the single most important factor in determining overall accuracy of the conductivity monitor is temperature compensation. Since temperature affects each solution conductivity differently, manual and automatic temperature compensators must match the process temperature response. By limiting the range of the ATC, closer matching and higher accuracies can be obtained. A slope adjustment control sets the slope of the ATC to match the process. Figure 4.12 shows a typical resistor–thermistor combination network. One or more of the fixed resistors can be replaced with a potentiometer to provide a slope or ATC calibrate control.

Modern conductivity systems provide accuracies to within 1% and stabilities as good as 0.05% of range per month with digital circuitry. Analog and digital meter readouts are available from most manufacturers. Typical response time is 1 s for a 90% input step change. Full-scale conductivity ranges vary from 0 to 1 μS to 0 to 1 S, with most monitors having built-in, isolated contact alarm relays available that are integral to the main circuitry.

The conductivity cell constant is the ratio of electrode spacing in centimeters to the electrode area in square centimeters and is given as cm^{-1}. Manufacturer's literature often eliminate the cm^{-1} unit designation.

Table 4.6. **Affect of Temperature on Conductivity: Ratio of Conductivity at Temperature Shown to Conductivity at 25°C**

Solution	Temperature, °C				
	0	25	50	75	100
Ultrapure water	0.22	1.00	3.11	7.46	14.20
5% NaOH	0.57	1.00	1.43	1.87	2.32
Dilute NH_3	0.50	1.00	1.47	1.83	2.05
Dilute HNO_3	0.65	1.00	1.31	1.58	1.80
Sugar syrup	0.34	1.00	2.41	4.40	6.93
10% HCl	0.64	1.00	1.33	1.63	1.87

Source: Courtesy of Beckman Instruments, Cedar Grove Operations.

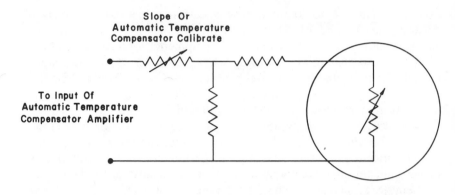

Figure 4.12. Typical automatic temperature compensation circuit. Material supplied by Beckman Instruments, Inc., Cedar Grove Operations. Copyright 1981, Beckman Instruments, Inc.

Probes with small cell constants have large electrode area, and the electrodes are closely spaced. Electrode materials are nickel, platinized nickel, Monel®, gold, titanium, and high-density carbon. Cell constants range from 0.001 to 50 with gold foil or gold wire electrodes used for the smallest cell constants and high-density carbon or platinized nickel electrodes used for high cell constants. Carbon and titanium electrodes are not subject to electrochemical polarization. Probes are supplied with cables ranging from 7 to 20 ft in length, and a total cable length of up to 500 ft is possible, depending on circuit design. Ryton®-bodied probes are available for use at temperatures of up to 175°C and pressures of up to 500 psi. Guarded electrode designs that permit the use of longer cable lengths are available from some manufacturers.

Conductivity monitors or pH monitors can be surface, panel, or pipe stand mounted. Figure 4.13 shows a conductivity cell mounted in a 1 × 1 × 1-in. piping tee. When pipe tees are used, the tee should be positioned so that the electrodes are completely immersed in the sampled liquid and no air entrapment can occur. Any uncovered cell area would change the cell constant. A flow-through cell holder design aids in keeping the electrodes clean.

Special equipment in combination with specially designed conductivity monitors are used to measure steam purity, condensate purity, the effectiveness of rinse control, and the effectiveness of reverse osmosis systems. Larson-Lane steam and condensate analyzers employ hydrogen ion exchange and degasification processes to detect variations in steam and condensate purity. Rinse control systems automatically sample the ionic

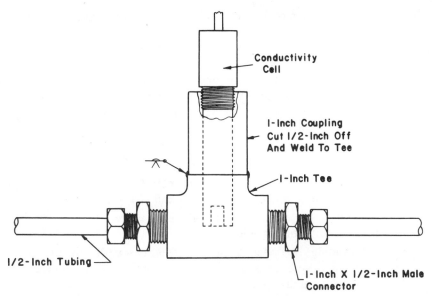

Figure 4.13. Direct insertion of conductivity probe in small pipe tee.

contaminate levels in rinse tanks and add potable or softened water until impurity levels are reduced to required preset levels. The conductivity setpoint is adjusted by means of a calibrated dial, and indicator lights indicate when the rinse bath is above or below the setpoints. Water cost savings of hundreds of dollars per tank per year are possible with these systems. Conductivity ratio monitors usually have dual readouts of conductivity ratio and percent rejection. In reverse osmosis (RO) systems a 90% rejection reading would indicate that 90% of the total dissolved solids in the incoming stream were removed from the outlet stream by the RO equipment. The conductivity ratio scale is used with large industrial cation exchangers or demineralizers, and the percent rejection scale (typically 99 to 80%) is used with RO and desalination systems. The ratio can be used to determine when to initiate regeneration of the beds, and the percent rejection scale can be used to determine when to backflush the RO system. Conductivity ratios of 0.5 to 1.5, 0.5 to 3.0, 0.1 to 1.1, 0 to 20, and 0 to 200 are available to cover most applications. A digital or analog meter is provided with the ratio controller.

The conductivity of HCl increases to a maximum at about 19 to 20% and then reverses. A similar reversal takes place with H_2SO_4 at about 36%. Naturally, conductivity measurements would be unacceptable as a monitor for acid concentration at or near these reversal points. There is, however, a wide range of acceptable use below and in some cases above these reversal points.

Other applications for conductivity monitors include:

1. Monitoring boiler feedwater and return condensate purity and controlling boiler blowdown to reduce the level of dissolved solids.

2. Monitoring cooling tower conductivity to control evaporation losses and needless increases in dissolved solids.

3. Detecting the leakage of corrosive materials through process heat-exchanger gaskets, seals, or defective tubing.

4. Monitoring wastewater, salt seepage from natural brine springs, and tidal saline intrusion into wells and streams.

5. Monitoring the effectiveness of distilled water and demineralized water production facilities.

6. Determining the concentration of acids, bases, or salts in various chemical processes such as pickling baths, caustic degreasing baths, and anodizing solutions.

7. Monitoring various food processing operations such as caustic pretzel cooking, caustic fruit and vegetable peeling, and the control of salt content in foods.

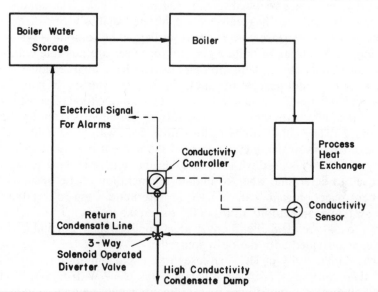

Figure 4.14.   Control of steam condensate return or dump based on conductivity measurement.

Figure 4.14 shows a conductivity monitor controlling a return condensate system. Although only one heat exchanger is shown in the figure, many heat exchangers can be regarded as being in parallel. Each heat exchanger in the system would have to have its own conductivity controller and three-way diverter valve. The conductivity sensor detects process leaks into the condensate line immediately downstream of the heat exchanger and dumps the potentially contaminated stream, thus protecting the return condensate line, boiler water storage tank, and boiler. Simultaneously a CCR alarm would be activated, pinpointing the leaking heat exchanger and warning the control room technician to take further action such as isolating the faulty heat exchanger.

4.2.4. General Comments

The greatest single factor affecting accuracy and reproducibility of a conductivity monitor is the ATC. The ATC must accurately sense the temperature of the solution, and the rate of change of the ATC circuit must precisely match the temperature–conductivity characteristics of the solution. Because of their improved resolution, digital panel meters tend to be more accurate and repeatable than analog meters, which were formerly widely used as display devices in control rooms or on the front panel of the monitor.

Conductivity monitors need not be used in closed-loop control to be effective. Process material leakage into a cooling water system often exhibits a go, no-go characteristic because the conductivity monitor would normally have a low on-scale reading but would be driven upscale when a process leak occurs. This on–off action could be used to open or close a valve or activate an alarm.

Future developments of exotic alloy electrodes will provide cells totally insensitive to electrochemical polarization. The incorporation of solid-state two-mode controllers and microprocessors has already occurred and has greatly increased the applications for these analyzers.

CHAPTER

5

MOISTURE AND
CORROSION MONITORS

Moisture monitors are used to detect and measure the concentration of water in solids, liquids, and gases. The amount of moisture in anhydrous acidic gases such as HCl or HF may be directly related to the corrosion rate of the associated process piping and vessels. Moisture also affects the product yield and catalyst life in catalytic reforming, the quality of chemical feedstocks, and the optimization of drying equipment.

Electrolytic hygrometers are frequently used to measure trace quantities of moisture in clean but corrosive gas streams. They can also be used to indirectly measure moisture in liquids when a stripping column with a dry gas stream is used to pick up the water from the sampled liquid and carry it with the gas to the hygrometer. There are two types of electrolytic hygrometers commonly used; one operates at low sample flow rates (typically 100 cm^3/min) and electrolyzes all the water in the sample, and the other operates at a higher flow rate but establishes an equilibrium condition where the electrolysis rate is proportional to the moisture content of the sample system stream. With this hygrometer a constant flow rate above 2000 cm^3/min is required. Both types of electrolytic hygrometer absorb and electrolyze the water present in the sample system in accordance with Faraday's valence law. Electrolytic hygrometers are low-range or trace moisture analyzers for use in gas streams with a maximum moisture content of 1000 ppm by volume.

Another moisture monitor widely used in industry is the impedance-type hygrometer with an aluminum oxide sensor. These wide-range hygrometers measure the water vapor pressure of the sample. They can be used to measure moisture in noncorrosive gas streams and low conductance, nonpolar liquids that follow Henry's law;[1] the concentration of a dissolved gas is directly proportional to the vapor pressure of that gas when in equilibrium with the liquid.

A third and relatively new type of industrial hygrometer utilizes a hygroscopically coated quartz crystal and oscillator circuit to measure moisture. This hygrometer is useful as a trend monitor for moisture measurements of up to 10,000 ppm by volume, and when equipped with an

88

optional microprocessor controller, a 25-fold increase in sensitivity and 20-fold increase in accuracy is possible for low-range applications. An internal moisture calibrator generates a known moisture standard for accurate calibration.

Moisture can usually be directly related to corrosion in industrial processes, but there are other factors such as process chemistry and piping metallurgy that may also influence corrosion rates. However, if moisture is a prime factor affecting corrosion, it may be desirable to install corrosion-monitoring equipment in the process lines or vessels.

There are two types of corrosion monitor commonly employed by industry. The first type was originally known as the linear polarization resistance (LPR) corrosion monitor. This type is now referred to as the *instantaneous corrosion rate* (ICR) monitor and is used only in polar or highly conductive liquids. Many ICR monitors can also be used to measure solution conductivity or externally applied millivolts. The second type is for nonpolar liquids and gases. Electrical resistance (ER) corrosion monitors are a relatively new development whose accuracy depends on precise temperature compensation and circuit stability. Probes for ER corrosion monitors can be considered as "automated corrosion coupons." The ER corrosion rate reading is not instantaneous; it must be calculated over a specific time period, which is usually 8 h or longer. As the ER probe corrodes, its resistance increases and high-gain circuitry is used to detect and measure this extremely small change in probe resistance.

5.1. MOISTURE MONITORS

Trace amounts of water are difficult to monitor because of the nature of water and the large amount of moisture in the atmosphere. Water tends to cling to every surface and accumulate in every crack and low spot in a system. Solid pipe and vessel walls as well as elastomeric materials seem to be able to absorb large quantities of moisture. Moisture can adversely affect feedstock and final product quality and in many cases considerably reduce operating life expectancy in process plants. For these reasons moisture monitors are considered important on-line analyzer systems.

The electrolytic hygrometer is simple, easy to apply and maintain, and generally accurate to within 5% of full scale on all instrument ranges. The accuracy of this type of monitor is dependent on the accuracy of the sample flow control system. This is covered in more detail later in this chapter.

Other factors affecting accuracy are:

1. *Recombination effects.* Samples having high concentrations of oxygen and hydrogen at low moisture levels exhibit higher than normal readings because the oxygen and hydrogen tend to recombine into water that is subsequently reelectrolyzed.
2. *Detector memory.* Detector cells contaminated with particulate matter can exhibit a memory for signal polarity. When this happens, an incorrect output signal will be obtained if the cell electrodes are reversed.
3. *Loss of sensitivity.* At extremely low moisture levels the cell coating may elute, and at very high moisture levels it may be washed away.

Despite these limitations the electrolytic hygrometer is considered to be one of the most accurate methods available for measuring low-level moisture in gases.

Impedance-type hygrometer probes are designed for direct insertion into process piping, thereby eliminating most sample conditioning requirements. They are used primarily in clean, noncorrosive gas service and in liquid streams that behave in accordance with Henry's law. The impedance-type hygrometer is not suited for use in highly conductive, polar solutions or most alcohol solutions.

The crystal oscillator hygrometer cannot be used at all with liquid streams or gas streams that tend to coat the hygroscopic surfaces of the crystal. This system is also more complex because of its sample system requirements and is thus more costly than the two systems previously mentioned. Depending on the application, sample system components such as internal or external gas dryers, containment traps, heated pressure reducing valve, or internal moisture calibrator are generally required.

However, the concept has been used on a worldwide basis for the past decade, and its reliability has been proved in the field. Analyzers of this type have been used to optimize catalytic reforming operations, drying operations in ethylene-producing units, and control moisture in natural gas–liquid separation plants.

5.1.1. Electrolytic Hygrometers

One type of electrolytic hygrometer sensor consists of two platinum or rhodium wires spirally wound on a small-diameter glass tube. Another version of the cell uses a U-shaped glass tube so that sample inlet and outlet are on the same end of the cell and a shorter cell length can be used. Figure 5.1 shows a cell designed for a flow rate of 2000 cm^3/min. In this

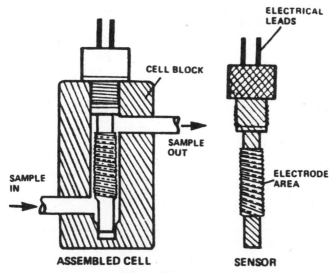

Figure 5.1. High-flow P_2O_5 electrolytic hygrometer cell. Courtesy of Ionics, Inc.

case the electrode wires are spirally wound on a small grooved cylinder that is threaded to fit inside a Du Pont Lucite® acrylic resin cell block. The surface area between the electrodes is coated with phosphorous pentoxide (P_2O_5). The response time of the cell is dependent on the thickness of the P_2O_5 coating; the thinner the coating, the faster the cell response time. In most cases the response time of the cell will be fast compared to the residence time of the sample system. The absorption of moisture from the gas stream on a film of P_2O_5 and the electrolysis of water react in accordance with the equations:

$$P_2O_5 + 3H_2O \rightarrow 2H_3PO_4 \qquad (5.1)$$

$$2H_3PO_4 \rightarrow P_2O_5 + 3H_2 + 1\tfrac{1}{2}O_2 \qquad (5.2)$$
$$\text{(during electrolysis)}$$

Oxygen and hydrogen are constantly being formed, and the P_2O_5 desiccant is being redried by the electrolysis current. When the sample flow is held constant, electrolysis current is directly proportional to the moisture content of the sample stream.

Figure 5.2 shows the circuitry of a typical electrolytic hygrometer. Cell voltage can be anywhere from 10 to 75 V DC. Range resistors are selected on the basis of cell operating voltage. In the circuit shown in Figure 5.2

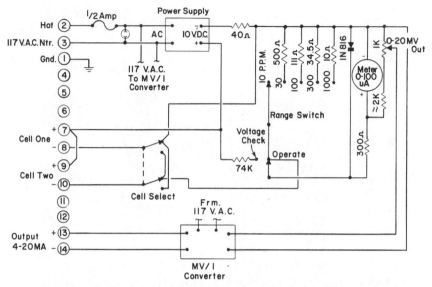

Figure 5.2. Simple electrolytic hygrometer circuit for dual-probe arrangement.

one of two electrolytic cells can be switched into the circuit. Dry nitrogen is used to keep the spare cell reasonably dry while the other cell is in service. In practice the power supply is tested by placing the "operate" switch in the "power check" position. The meter needle should go to a premarked "power good" position. Then the operate switch is returned to the "operate" position and the range switch is placed in a range position setting where the meter is on-scale. After the cell has reached equilibrium with the moisture in the process, a range switch position is chosen which produces a low, on-scale meter reading. The 4- to 20-mA output signal is used to provide a computer or recorder trend trace and high-moisture alarm if desired. The mV/I converter is calibrated so that a 0- to 20-mV input produces a 4- to 20-mA output signal.

Other electrolytic hygrometer circuits incorporate the use of an amplifier with zero, gain, and span controls for calibrating the instrument over a wide range of operating conditions.

Figure 5.3 shows the electrolysis current as a function of moisture content and sample flow rate. The curves relate to the cell shown in Figure 5.1 when it is powered by a 75-V DC supply. Sample flows of 2000 cm^3/min should be maintained to maximize accuracy since the curves are relatively flat at this setting.

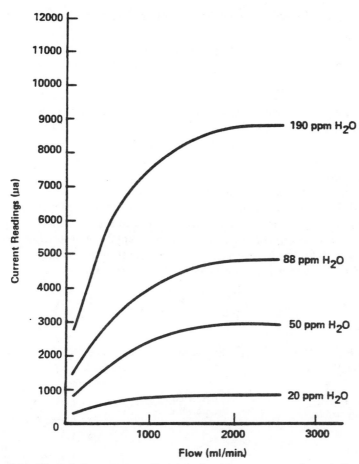

Figure 5.3. Electrolysis current as a function of moisture content and sample flow for the high-flow P_2O_5 sensor. Courtesy of Ionics, Inc.

5.1.2. Aluminum Oxide Hygrometers

The aluminum oxide or impedance-type hygrometer was originally designed for the aerospace industry, where severe environmental conditions were encountered. It was quickly adopted by industry as an *in situ* water vapor monitor for process applications because of its ruggedness.

The basic sensor consists of an aluminum strip with one side anodized to produce a porous oxide layer. A thin layer of gold is then vapor deposited on the anodized surface. The aluminum base metal and the thin gold layer form the two electrodes of an aluminum oxide capacitor.

Water vapor is rapidly transported through the porous gold layer and equilibrated on the pore walls in direct proportion to the partial pressure of water in the vicinity of the probe. As water molecules are absorbed by the sensor, the electrical conductivity of the pore wall increases, thereby decreasing the probe impedance. Water is selectively admitted to the pore wall by control of the size of pore wall openings. Water molecules are small compared to the size of organic molecules, so only water permeates the porous gold layer. The aluminum oxide sensor design is shown in Figure 5.4.

The aluminum oxide sensor is used as a direct insertion probe in both gas and liquid service. It is unaffected by ambient temperature and pressure and the hygrometer can be calibrated in terms of partial pressure of water in millimeters of mercury or dew point temperature in degrees Celsius or Fahrenheit. The volumetric concentration of water in a gas is related to the vapor pressure of water by the equation:

$$\mathrm{ppm}_v = \frac{P_{\mathrm{H_2O}}}{P_t} \times 10^6 \qquad (5.3)$$

where $P_{\mathrm{H_2O}}$ is the partial pressure of water as measured by the hygrometer

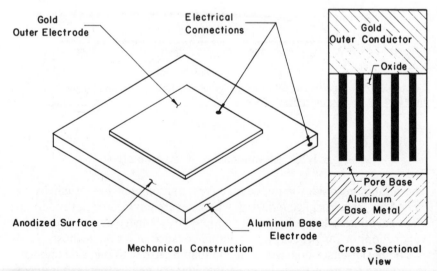

Figure 5.4. Construction of the aluminum oxide hygrometer. Courtesy of Panametrics, Inc.

and P_t is the total pressure of the system. If the total system pressure increases, the vapor pressure of the water also increases. Each probe must have its own specific calibration curve because of minute differences in them as they are manufactured.

In gas service the term "percent relative humidity" (%RH) can be calculated from the equation:

$$\%RH = \frac{P_{H_2O}}{P_{sat}} \times 100 \tag{5.4}$$

where P_{sat} represents the saturation vapor pressure of water at a given temperature.

The aluminum oxide sensor has the highest sensitivity and greatest dynamic range of moisture measurement in gases. Its ability to measure dew point at pressures from a few microns of mercury to 5000 psig is a great advantage because the absolute amount of water present in the sample can be determined at the actual operating pressure of the sample. Operation at high pressure results in increased sensitivity, better accuracy, and faster response time.

Equation (5.4) is used to determine the percent saturation of water in a liquid. The concentration of water in a liquid is expressed in terms of weight percent or ppm_w. The amount of water that can be absorbed by a liquid is related to Henry's law constant by the equation:

$$ppm_w = K \times P_{H_2O}$$

where K is Henry's law constant and P_{H_2O} is the vapor pressure of water as measured by the hygrometer.

Liquids vary widely in their ability to absorb moisture and their saturation values change dramatically. For example, at 20°C the saturation value of benzene is 639 ppm_w and of cyclohexane, 122 ppm_w. A conversion table for dew point, vapor pressure, percent relative humidity, and volumetric and weight concentration is presented in Appendix 9. Moisture levels in liquids in the order of 5×10^6 ppm_w can be determined. This if four orders of magnitude better than Karl Fischer procedures and three orders of magnitude lower than other more complex methods.

5.1.3. Crystal Oscillator Hygrometers

The crystal oscillator hygrometer measures moisture by monitoring the frequency change of an oscillator by use of a hygroscopically coated quartz crystal that is alternately exposed to moist and dry gas. A sealed reference crystal provides a constant frequency for comparison. Both crystals are temperature controlled at 60°C to assure good repeatability and accuracy.

The sample system of the hygrometer shown in Figure 5.5 divides the incoming sample stream. One stream is carried to the measuring crystal, and the other stream is dried before being routed to the measuring crystal. The dried stream is the reference gas stream of the analyzer. The two streams are alternately switched every 30 s, thus exposing the measuring crystal alternately to moist and dry sample streams.

As moist sample gas enters the crystal, cell moisture is absorbed by the hygroscopic coating of the crystal, causing a change in oscillator frequency. At the end of the 30-s period the microprocessor reads the oscillator frequency and switches gas flow so that dry sample gas is passed to the crystal cell for the next 30s. At the end of this 30-s period the microprocessor again reads the oscillator frequency. The two frequencies are then electronically compared to the frequency of the uncoated sealed reference crystal, and the moisture level is computed, linearized, and displayed on a front panel digital meter. A binary-coded decimal (BCD) output or optional linear analog output signal with track and hold feature can be provided. The analog output signal is updated every 30 s when the track and hold feature is provided.

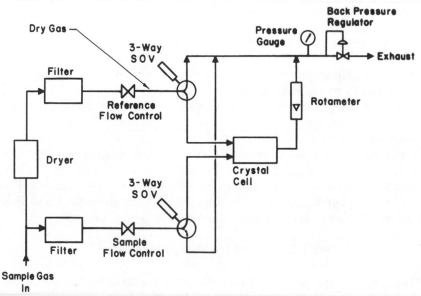

Figure 5.5. Sample system for the crystal oscillator hygrometer. Courtesy of DuPont Analytical Instruments.

When the optional internal gas calibrator is used, the dry gas stream is also divided and a portion of it passed to the moisture generator. The moisture generator is temperature controlled at 60°C and consists of a permeable Du Pont TFE fluorocarbon polymer coil and water-saturated wicking arrangement. In the calibrate mode the analyzer alternately compares the crystal frequency of dry gas to that of the wet gas produced by the moisture generator.

5.1.4. Calibration Techniques

The calibration of the low-flow (100-cm^3/min) electrolytic hygrometer is considered to be absolute. The cell either responds properly or does not. There are no zero or span controls, so the indicating meter circuit is designed to provide a meter reading corrected for a sample of 60°C with a pressure of 25 psig. Since water merely equilibrates on the desiccant of the high flow instrument and is not quantitatively electrolyzed, this hygrometer has zero and gain controls for calibration.

Electrolytic hygrometers are often difficult to calibrate in the field. Sometimes a comparison can be made with laboratory results, or laboratory-type dew–frost point equipment can be temporarily used in the field. At other times it may be easier to use another electrolytic hygrometer with a newly sensitized cell in series with the suspect instrument. Electrolytic cells are resensitized by rinsing them with acetone, then distilled water, and acetone again. After drying, a 25 to 45% aqueous solution of phosphoric acid is flushed through the cell several times to remove particulate matter and completely wet the cell surface. The cell is slowly rotated under a heat lamp to evaporate most of the water and then purged with a low flow of nitrogen with the cell power turned on. When a test hygrometer reading indicates that the cell is dry, it is ready for service.

Sample system lines should be kept as short as possible, and sample system volume should be kept small. The sample system shown in Figure 5.6 is suitable for use with both the low- and high-flow (2000-cm^3/min) electrolytic hygrometers. The bypass rotameter is used to increase incoming sample flow for reduced sample transport time. The filter on the inlet line provides protection from small conductive particles that can short-circuit the electrodes of cells operating at 75 V DC, resulting in permanent cell damage. When the electrolytic hygrometer is not monitoring the sample stream, the instrument should be purged with reasonably dry (<10 ppm$_v$) nitrogen gas. The air heater shown in Figure 5.6 is used to heat the sample system components to help maintain the sample temperature at 60°C.

Aluminum oxide hygrometers are usually mounted directly in the

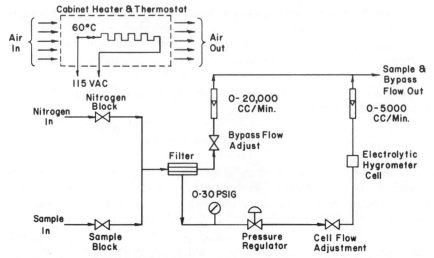

Figure 5.6.   Sample system for the P_2O_5 electrolytic hygrometer. Courtesy of Ionics, Inc.

stream to be monitored. There are few sample system requirements except for the fact that liquid samples must be temperature controlled because the constant in Henry's law varies with temperature. Impedance-type hygrometers have a built-in "Cal" switch or pushbutton. The calibrate switch checks for proper oscillator voltage, admittance amplifier calibration, log amplifier linearity, and output circuit response. It does not test the probe.

If the process medium is normally dry, a quick check of probe response can be made by removing the probe and exposing it to ambient air. A positive dew point reading should be observed within 5 s. A similar dry-down response should be noted when the probe is reinstalled in the process. If moisture readings are less than 1000 ppm_v at atmospheric conditions, it may be possible to check the sample stream with a portable electrolytic hygrometer.

Calibration of the crystal oscillator hygrometer can easily be verified by using the optional factory installed and adjusted moist gas generator. A controlled amount of moisture is added to the dry gas stream and then routed by way of solenoid valves to the detector cell, where a stable reading is generally obtained in a few minutes. A continuous flow of gas through the calibrator keeps it in a ready state when it is not being used for detector calibration.

Samples of gas and liquid streams may also be taken to the analytical laboratory for analysis. When laboratory results are used for comparison purposes, it should be kept in mind that sampling techniques and the nature of airborne water could contribute greatly to inaccuracies.

5.1.5. Design Features and Applications

The electrolytic (P_2O_5) hygrometer uses a direct current power supply for cell excitation voltages, which can vary from 10 to 75 V depending on the application. When a second cell is used, it is usually placed into the sample system as an installed spare and kept purged with dry nitrogen. The electrolytic hygrometer is small and simple and lends itself to a variety of portable designs. Small porous filters are usually installed in the sample inlet line to protect the flow control devices and small aperature sensor cell because the electrode wires are spaced only $\frac{1}{32}$ in. apart. The sensor must. be protected from conductive particles that might short-circuit the electrodes and permanently damage the cell. This monitor is not recommended for oxygen- or hydrogen-rich streams, alcohols, amines, ammonia, or hydrogen fluoride streams. Alcohols are seen as water, and some of the other chemicals attack the P_2O_5 desiccant.

Electrolytic hygrometers are usually multirange instruments having typical moisture ranges of 0 to 10, 30, 100, 300, and 1000 ppm_v. The accuracy of the electrolytic hygrometer is within $\pm 5\%$ of the measurement range used. Sensitivity decreases because of the gradual loss of the P_2O_5 particles that form the desiccant coating on the cell. The service life of the cell can vary, but cell coating lives of several weeks or months are typical. Time response of these hygrometers is somewhat slow, usually several minutes, even in a sample loop of only 500 cm^3. This slowness of response is not necessarily a handicap because moisture levels in process streams usually tend to change gradually over a matter of hours.

Aluminum oxide hygrometers typically produce a 63% detector response to a 100% input change in only 2 s. The normal operating range of this hygrometer is -100 to $+20°C$ dew point, corresponding to 0.001 to 20,000 ppm_v. Operation to a $+60°C$ dew point or 200,000 ppm_v is possible at pressures from a few microns or mercury to 5000 psig. Many units feature a multiple probe arrangement capable of handling three to six sensors with input signal cable lengths of up to 3000 ft. These probes cannot be used with materials that will corrode aluminum or plug the pore wall openings of the probe by polymerizing on contact with the probe surfaces.

Crystal oscillator hygrometers are now available equipped with internal moisture generators for calibration verification and microprocessor control of calculation and data display functions. They are relatively easy to maintain, and the microprocessors often have a diagnostic program capability to help troubleshoot hygrometer malfunctions. These hygrometers will respond within 3 min to a two-decade step change in the moisture content of the sample.

Crystal oscillator hygrometers have a measuring range of 0 to 9999 ppm_v. Input spans as narrow as 0 to 5 ppm_v are available. They are relatively specific to water but cannot be used with liquids and should not be used with materials that will attack the hygroscopic crystal coating or polymerize on it. Accuracy is generally better than ±10% of the measuring span used. Standard measurement spans are 0 to 25, 0 to 50, 0 to 250, and 0 to 1000 ppm_v. A sample flow of 250 cm^3/min at 15 psig should be provided for the sensor. Maximum sample pressure should not exceed 100 psig. The moisture sensor may be separated as far as 2000 ft from the hygrometer mainframe.

Applications for moisture monitors include measuring the moisture content of instrument air, process oxygen, purge nitrogen, natural or process gas, and most refrigerants. Moisture in natural gas from producing or storage wells adversely affects the BTU rating of the gas and can cause pressure reducing equipment to freeze. Very small quantities of water in hydrocarbon streams being separated by cryogenic fractionation and distillation can cause hydrate formation, leading to equipment plugs and process blockage. Moisture monitors are often incorporated into process drier valve switching and refrigeration controls to automatically change dryers as required based on analysis of the drier effluent gas moisture content. The moisture content of blanketing gas and exothermic base gases should also be measured and controlled because moisture can cause undesirable oxidation of some of the key components in annealing oven and heat-treating processes.

Humidity must be controlled in bag house blower air because high humidity can clog filter bags, thereby greatly reducing their efficiency. More recently the monitoring of flue gas moisture has taken on new significance in optimizing combustion efficiency. Maintaining moisture levels high enough to assure complete combustion of combustible gases helps satisfy governmental regulations concerning pollution. Moisture is directly related to combustion efficiency because it is one of the products of complete combustion. When moisture levels fall below a predetermined level, combustion airflow is usually increased.

Other applications include:

1. Monitoring air humidity in electronic equipment rooms, simulated space chambers, and radar waveguides.

2. Monitoring the moisture content in fluorocarbon or other highly reactive gases used in acidic processes susceptible to corrosion.

3. Monitoring moisture of annealing furnace atmospheres and blanketing or padding gas.

The hygrometer best suited for a given application depends on many variables such as the corrosive properties of the material being monitored, the expected moisture range, and the sample system pressure and temperature requirements.

5.1.6. General Comments

Infrared absorption, heat of adsorption, microwave absorption, and neutron moderation hygrometers have not been covered in this chapter. Infrared absorption analyzers are covered in Chapter 7. The heat-of-adsorption hygrometer requires extensive sample conditioning and is less commonly applied in industry, and its sensor cannot be exposed to liquids. Microwave absorption and neutron moderation hygrometers are normally used to measure moisture in pastes, slurries, and solids.

All the hygrometers commonly employed by process industries to measure moisture in liquids and gases require the sample stream to be reasonably clean and noncorrosive for satisfactory sensor life. The chemical resistance of the hygroscopic desiccant coatings used on the sensors must also be considered.

Electrolytic and aluminum oxide hydrometers will probably continue to be popular because they are simpler in design, require fewer sample system components, and are generally more economical than other more complex moisture-measuring systems currently available. In the future there will be an increased use of digital data displays and microprocessors to increase accuracy, enhance versatility, and reduce equipment maintenance.

5.2. CORROSION MONITORS

Corrosion—like old age—is a natural, continuing process that can be slowed down but not stopped. Its annual cost to the process industries exceeds a billion dollars. Reduction of corrosion rates to their lowest possible levels minimizes this cost by extending equipment longevity and helping conserve energy, natural resources, and maintenance labor. Corrosion monitors are most often used to determine corrosion rates in terms of mils per year (mpy), where a mil is defined as $\frac{1}{1000}$ inch. Corrosion monitors are most accurate and effective when monitoring uniform or general corrosion and least effective in monitoring pitting or localized corrosion.

Table 5.1 is a comparison of some of the corrosion monitoring techniques used by industry. Eddy current and ultrasonic testing techniques are also utilized to detect cracks, flaws, and other imperfections in process piping and vessel walls because these areas are often prone to corrosive attack.

5.2.1. Linear Polarization Resistance Monitors

During the 1950's A. L. Geary and Milton Stern did development work on the measurement of instantaneous corrosion rate by means of polarization data.[2,3] Stern's method made use of a parameter known as *polarization resistance,* which is the slope of the polarization curve in the region near the corrosion potential. Polarization resistance is defined as:

$$R_p = \frac{dE}{dI} = \frac{B_a \times B_c}{2.3\ (I_c)(B_a + B_c)} \tag{5.5}$$

where dE/dI is the change in electrode potential proportional to the change in electrolyte current, B_a and B_c are constants representing the slope of local logarithmic anodic and cathodic polarization curves, and I_c is corrosion current. Figure 5.7 shows an ideal curve for polarization resistance as a function of corrosion current where both B constants are 0.1.

Stern's work also pointed out that values for the B constants typically ranged from 0.06 to 0.12 and that the corrosion rate could be determined to within an accuracy of 40% without any knowledge of B values. He concluded that corrosion current was a good indication of corrosion rate when the current was adjusted to produce a small but constant electrode polarization voltage of about 5 to 10 mV. This principle is illustrated in Figure 5.8 with the use of three electrodes, an ammeter, a voltmeter, and an adjustable current source. The meters have adjustable null features to compensate for normal variations in the surface conditions of the test and reference electrodes. A small electrical current flows between the test and auxiliary electrodes. Since corrosion occurs at the anode the test electrode is protected when it is the cathode. The direction of current flow is then reversed, and the test electrode becomes the anode and its corrosion rate is accelerated. The change in corrosion rate caused by a reversal in current causes a change in potential of the test electrode when compared to the freely corroding reference electrode. The relationship between the

Table 5.1. Corrosion Monitoring Techniques

Method	Area of Application	Type of Data Produced	Limitation of Method
Weight-loss coupons	All processing industries	Long-term performance data on general corrosion; good for some forms of localized corrosion	Integrates data for calculating average corrosion rate over the exposure period; high workforce requirements to perform coupon inspection; usually only during scheduled unit downtime
Linear polarization corrosion rate	Electrolyte systems	Real-time operational data on general corrosion; readout in mils per year	Electrolyte such as a highly conductive aqueous solution typically required
Electrical resistance	All processing industries	Medium- to long-term data on general corrosion or erosion	Accumulates data between readings; computer or computing module calculates mils per year
Potentiodynamic polarization	Electrolyte systems	General and localized corrosion	Requires expert interpretation of potentiodynamic charts

Source: Adapted from information supplied by Rohrback Instruments.

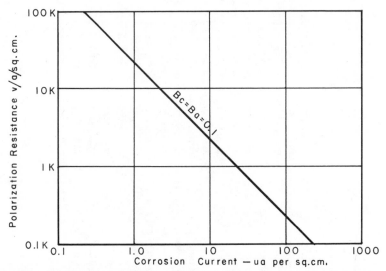

Figure 5.7. Polarization resistance as a function of corrosion current; an important relationship in the development of LPR corrosion monitors.

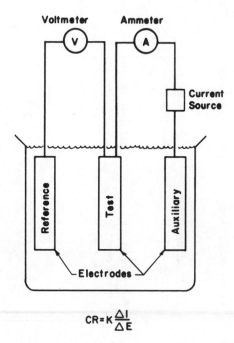

$$CR = K \frac{\Delta I}{\Delta E}$$

Figure 5.8. Basic diagram of the three electrode LPR corrosion monitor. Courtesy of Petrolite Instruments.

104

change in current flow dI, the change in polarization voltage dE, and the corrosion rate CR of the test electrode is given by the equation:

$$CR = K \frac{dI}{dE} \tag{5.6}$$

Since all corrosion rate measurements are made at a constant polarization potential of typically 10 mV, the voltage term becomes constant, so the corrosion equation can be reduced to:

$$CR = KI \tag{5.7}$$

As long as the polarization voltage is constant, the corrosion rate is directly proportional to the current required to produce that polarization voltage. If the corrosion rate is low, a relatively low current is required; if the corrosion rate is high, a relatively high current is required. A value representing the K factor is designed into the instrument so that readings are linear and calibrated directly in mils per year.

Both LPR or ICR instruments are unique in that they measure the instantaneous corrosion rate instead of the average corrosion rate. A graph of instantaneous corrosion rate plotted against time is shown in Figure 5.9. The average corrosion rate is shown by the dashed horizontal line. Corrosion rates of pipe and vessel systems can typically vary by one

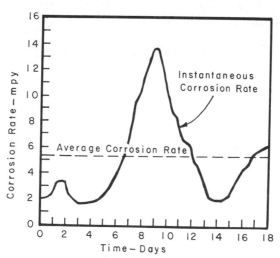

Figure 5.9. Instantaneous corrosion rate versus time. Note that the average corrosion rate provides little useful information on a day-to-day basis.

or even two orders of magnitude. These variations in corrosion rate are extremely difficult to detect by corrosion coupon or other corrosion averaging technique. Both LPR and ICR instruments produce quick answers, enabling process engineers to study corroding system dynamics in real time. It no longer takes weeks or months to observe the effect of adding various corrosion inhibitors to the system.

Corrosion probes should remain installed throughout the duration of a test period so that the surfaces of the test electrode can react the same way as the pipe and vessel walls in a system. Electrode equilibrating time can vary from a few hours to as long as a week before corrosion monitor readings are representative of the corrosion rate of other metal surfaces in the system. However, LPR probes will not work well in liquids that have specific resistivities greater than 10 megohm-cm or pure hydrocarbons at normal temperatures.[4] Also, their effectiveness in detecting pitting and certain other types of localized corrosion varies widely depending on application.[5] Manufacturers presently supply both two- and three-electrode LPR probes. The two-electrode probes are suitable for use where the solution resistivity in ohm-cm and mils per year product is 5000 or less. In solutions that have a higher multiplication product the three electrode system is more accurate.[6,7]

Figure 5.10 shows a standard probe assembly with cylindrical removable electrodes that can also be used as corrosion coupons. The electrodes are weighed before and after exposure to the process to obtain weight loss and average corrosion rate data. The total weight loss of the coupon during the exposure period provides the data used to extrapolate the annual average corrosion rate.

5.2.2. Electrical Resistance Monitors

The electrical resistance corrosion monitor was first developed in the late 1950's by oil refinery engineers. The monitor basically consists of a low-resistance Wheatstone bridge circuit with one side of the bridge containing the measuring element and the other side having a reference element. Figure 5.11 is a simplified diagram of an ER corrosion monitor showing a probe having an exposed measuring element and protected reference element, a Wheatstone bridge measuring and output circuit, and a power supply. The measuring element is a loop of wire that is exposed to the corrosive environment. The reference element is encapsulated inside the probe body with thermally conductive plastic so that it can sense the process temperature even though it is not directly exposed. The output meter indication generated by an ER monitor is directly proportional to the resistance of the measuring element. Corrosion causes the thickness

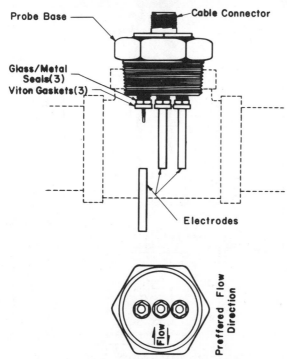

Figure 5.10. Direct insertion LPR probe with replaceable electrodes. Electrodes can also be used as corrosion coupons. Courtesy of Petrolite Instruments.

of the measuring element to decrease, resulting in an increase of its resistance.

Figure 5.12 shows a block diagram of an electromechanical ER corrosion monitor. A square-wave oscillator is used to drive the Wheatstone bridge and servomotor system. The square-wave oscillator also provides a reference voltage to the phase-sensitive detector to keep it synchronized with the power to the probe. The synchronized amplifier amplifies the reference voltage signal, and the phase-sensitive detector converts the amplified AC probe signal to DC and provides additional noise rejection. The output of the phase-sensitive detector is a differential DC voltage that is amplified and converted to single-ended DC voltage by the DC amplifier. An output transformer supplies 0.5 A to the probe so that probe resistance changes of only 100 to 1000 $\mu\Omega$ can produce input signals of 50 to 500 μV. These very small input signals require a high-gain, multiple-stage amplifier to produce suitable output voltages.

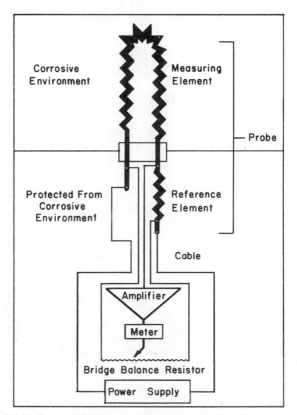

Figure 5.11. Basic diagram of ER corrosion monitor with wire loop probe. Courtesy of Rohrback Instruments.

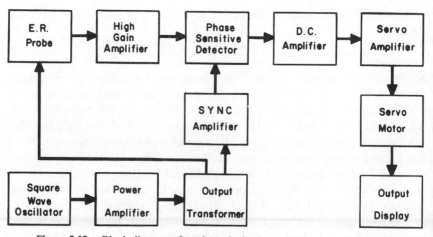

Figure 5.12. Block diagram of modern single-channel ER corrosion monitor.

108

The average probe wall thickness can be calculated from the following equation:

$$WT = \frac{KM}{1000 - M} \qquad (5.8)$$

where K is the initial wall thickness in mils and M represents the meter reading (0 to 1000). A decrease in probe wall thickness causes the output meter display to decrease. For example, a new probe with a 20-mil wall thickness might typically produce an output indication of 500, whereas a spent probe with only a 5-mil wall thickness remaining might show 200. This change in reading over a period of 3 months would infer an annual average corrosion rate of 60 mpy because there was a 15-mil wall thickness loss in $\frac{1}{4}$ yr.

Corrosion coupons often require long exposure to corrosive environments and usually require plant shutdowns for installation or removal. Highly qualified personnel and reasonably sophisticated test procedures are required for the interpretation of results.

Another advantage of the ER corrosion monitor over corrosion coupon monitoring is shown by the graphs in Figure 5.13. A corrosion coupon can

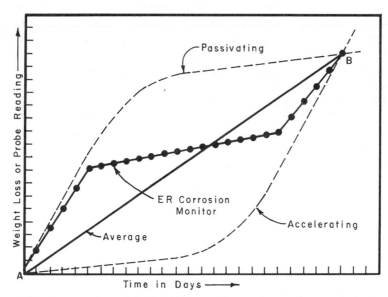

Figure 5.13. A comparison of simulated ER corrosion monitor readings to passivating and accelerating corrosion curves. Note that the average corrosion curve from points A to B does not represent day-to-day operating experience. Courtesy of Rohrback Instruments.

be used to determine only the average corrosion rate during an exposure period. Coupons cannot be used to determine if corrosion rates are passivating (decreasing) or accelerating as indicated by the dashed lines from points A to B. It is actually more likely that the corrosion rate changes constantly depending on process operating conditions, chemical activity, and ambient conditions as suggested by the changing slope on the trace marked "ER corrosion monitor." A computer can compare the existing ER probe reading to the initial probe reading or a historical reading at some other specified time and calculate the average corrosion rate over a given exposure period. In Figure 5.13, the data points show that the corrosion rate was calculated on a daily basis. Comparative corrosion readings corresponding to corrosion rates of less than 2 to 3 mpy are not significant. Even in rapidly corroding systems the minimum time period for corrosion rate calculations is typically 8 h.

Welded tubular probes with concentric measuring and reference elements are designed for operation in systems having temperatures as high as 1000°F and pressures up to 5000 psig.

5.2.3. Calibration Techniques

Meter provers or probe simulators are passive voltage dividers used for verifying instrument operation and cable integrity. The resistors are usually encapsulated in devices that are equipped with matching cable connectors allowing direct substitution of the corrosion probe. LPR-type simulators provide corrosion rate signals corresponding to 1, 2, 5, and 10 mpy.

ER corrosion monitor probe simulators can also be used to determine "good" and "bad" or "new" and "spent" probe responses. Typical ER probe resistance values for a 20-mil carbon steel probe are listed in the table shown on Figure 5.14. It should be noted that resistance values will vary widely depending on probe material, manufacturer's design, wall thickness, and other variables. The table in Figure 5.14 demonstrates that ER corrosion probes are extremely low resistance devices.

Checking a corrosion monitor involves little more than connecting the probe simulator to the mating connector on either the instrument or the field cable. The LPR monitor should provide an output corresponding to the probe or simulator value in mils per year. When a probe simulator is connected to an ER corrosion monitor, the output meter should indicate the equivalent simulator resistance. Some simulators have a dual-range switch that allows the circuitry to be checked at two different output values.

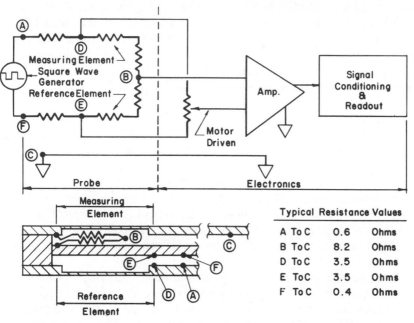

Figure 5.14. ER corrosion probe physical arrangement and wiring details.

5.2.4. Design Features and Applications

Fixed or retractable probes can be supplied for all LPR corrosion monitors. Retractable probes are usually mounted through full-opening ball valves. Probe operation at temperatures up to 150°C and pressures to 3000 psi is possible.

Most manufacturers of LPR corrosion monitors offer a number of portable and in-line models having selectable full-scale ranges of 1, 3, 10, 30, 100, 300, or 1000 mpy. Industrial-type in-line LPR corrosion monitors provide analog or digital readouts; anodic or cathodic polarization voltage modes; single or multiple probe inputs; computer compatibility; and switch selectable outputs of 0 to 4, 0 to 16, 0 to 40, .1 to 5, 4 to 20, or 10 to 50 mA. One multichannel unit features continuous active monitoring of up to six probes as well as a continuous output for each channel.[8] In-line LPR corrosion monitors can be calibrated for corrosion rate or conductivity ranges from 0 to 0.2 to 0 to 1000 mpy or mS. Cable lengths exceeding a mile are possible with most models.

ER corrosion probes with wire loop elements are usually provided with tubular shields to protect them from physical damage. Probe temperature ratings of up to 400°C and pressure ratings of up to 4000 psi are typical with wire loop elements of 40- or 80-mil diameters. Probes with tubular elements are available in 4- and 8-mil wall thicknesses. Probes usually are kept in service until there is only 25% remaining of the original wire diameter on wire loop probes or 50% remaining of the original wall thickness of the 4- and 8-mil tubular elements.

Some single-channel ER corrosion monitors have special circuits providing a digital display range from 0 to 1000 units with a resolution of 0.1% of the full-scale selected and sensitivity of 0.5 μin. per digit. A microprocessor-controlled, remote multiplexing ER corrosion monitor is available that can handle a maximum of 240 probes and communicate new data to a central computer system every 30 min.[9] A request from the central or host computer causes the microprocessor to transmit currently stored data using selectable communications formats of RS232, RS422, or a 20-mA current loop at rates up to 9600 baud. The host computer is often used to calculate system metal losses, probable remaining wall thicknesses, and annual corrosion rates.

Another microprocessor-controlled unit makes all critical calculations for a six-probe monitoring arrangement. The instrument sequentially reads each probe and calculates the current corrosion rate and totalized metal loss. Results are printed in engineering units, either in mils per year or millimeters per year. Corrosion rate alarm setpoints for each channel are entered through the front panel keyboard. Visual and audible alarms activate for invalid probe data, or when high short-term corrosion or probe life is exceeded. This unit has a battery for power backup and features relay contacts for the activation of remote alarm systems.

LPR corrosion monitors are used extensively in aqueous systems whose conductivity falls between 0.1 to 1000 mS. Industrial uses include oil field evaluations of inhibitors for the control of water, oxygen, CO_2, and brine. In these applications it is possible to determine:

1. The effectiveness of inhibitor formulas.
2. Dosage requirements as a function of inhibitor injection cycles.
3. Minimum inhibitor levels and costs.
4. The effectiveness of continuous versus batch injection treating.

Another use of LPR corrosion monitors is for checking water injection systems for the evaluation of biocides, oxygen scavengers, and filming inhibitors. Additional applications for these monitors include:

1. Corrosion monitoring of process equipment and connecting pipe systems.

2. Detection of process leaks into steam condensate return systems.
3. Determining the corrosion resistance of various metals and alloys in chemical plants.
4. Determining how pH variations, temperature, stream velocity, and contaminants affect the corrosion rate.
5. Studying passivation film buildup and anodic protection techniques.
6. Evaluating the effects of startup and shutdown procedures on corrosion rates.
7. Studying corrosion mechanisms in batch reactions in order to select the most cost-effective materials of construction.
8. Optimizing deaeration procedures, temperature, and feed stream composition to minimize corrosion in brine evaporators.
9. Monitoring sour gas stripping operations in natural gas production plants.
10. Studying container and content reactions in the food and canning industries.

The ER corrosion monitoring system can be used wherever corrosion might be a problem provided the temperature and pressure limits of the probe are not exceeded. ER probes are used extensively throughout systems in oil refineries and chemical processing plans.[10] They can also be used to monitor the affect of contaminated air on sensitive electronic circuits and check control room air cleaning systems.

The primary limitation of an ER corrosion monitor is that the instantaneous corrosion rate is not known, so any corrective action must be delayed until an average can be determined. This limitation is rapidly disappearing with the adaptation of microprocessor technology to these monitors. The next generation of ER corrosion monitors will be more flexible in application, more reliable, and more accurate because of the precise matching of probe characteristics with circuit functions.

5.2.5. General Comments

Corrosion causes about 70 billion dollars' worth of industrial damage each year. High corrosion rates in most processes are not inevitable. Corrosion is a variable that can be continuously monitored and controlled. The state of this art has now developed to the point where continuous in-line monitoring should be considered wherever corrosion problems exist.

CHAPTER

6

OXYGEN ANALYZERS

Oxygen analyzers are used to monitor oxygen levels in processes that manufacture hydrocarbons, the food processing industry, and the mining industries.

Oxygen analyzers have also been increasingly used as stack gas monitors for optimizing fuel: air ratios to achieve maximum burner efficiency in boilers and other oil-, gas-, and coal-fired heaters. Stack gas oxygen analyzers are frequently installed with combustion monitors to make determinations of fuel losses from the combustion process. Each 0.1% of combustibles in the flue gas represents a loss of 0.3 to 0.6% of the fuel used.

Dissolved oxygen analyzers have found increasing use in industrial wastewater aeration systems, activated sludge ponds, and municipal water treatment plants. They are also used to monitor the dissolved oxygen levels of lagoons, lakes, and streams. Dissolved oxygen levels must also be carefully controlled in boiler feedwater and many food and beverage processing applications.

Dissolved oxygen monitors using thallium cells have decreased in popularity because thallium is a heavy metal that slowly dissolves in water. Thallium is toxic to humans, and the symptoms of thallium poisoning are similar to those of lead poisoning. Manufacturers of thallium probes provide a warning to "wash hands after handling thallium probes and before smoking cigarettes."

6.1. GAS ANALYZERS

In situ stack gas analyzers are being more widely used now than older oxygen analyzers that require sample systems. Most of these *in situ*-type oxygen analyzers use zirconium oxide cells.[1] Therefore, special emphasis is placed on zirconium oxide cell operation in this section. Several modern variations of oxygen analyzers operating on this principal are covered in detail.

114

6.1.1. Zirconium Oxide Cell

The zirconium oxide ceramic sensing element is a closed-end tube that is heated to about 1500°F. At that temperature the ceramic becomes an electrolytic conductor as vacancies in the crystal lattice permit passage of the oxygen ions. The porous platinum coatings on the inside and outside of the ceramic tube act as electrodes. When these electrodes are in contact with gases having different levels of oxygen partial pressure, a voltage is produced that is proportional to the log of this ratio. Air is normally used as the reference gas on one side of the cell so that the potential produced by the cell is proportional to the oxygen content of the sample gas. Figure 6.1 shows cell output in millivolts as a function of percent oxygen concentration of the sample. The equation for calculating cell output is:

$$E = AT \log \frac{0.209}{O_2 x} \qquad (6.1)$$

where A is a constant of proportionality, T is absolute temperature in degrees Kelvin, and $O_2 x$ is the oxygen concentration of the sample.

Figure 6.2 shows the principle of operation of the zirconium oxide cell. A positive sample pressure of about 0.4 in. of water is needed for ade-

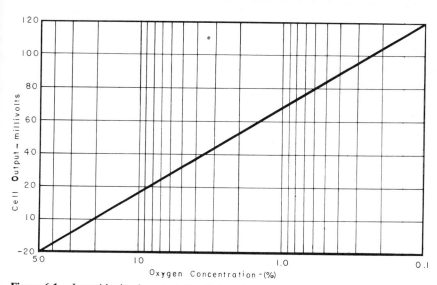

Figure 6.1. Logarithmic characteristic of zirconium oxide oxygen sensor. Courtesy of Ametek/Thermox Company.

OXYGEN ANALYZERS

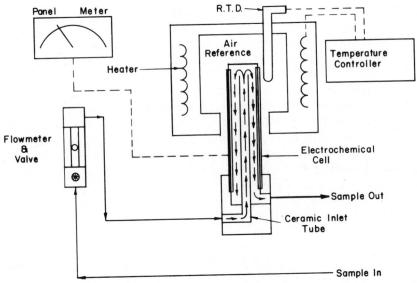

Figure 6.2. Zirconium oxide sensor and basic sample system conditioning arrangement. Courtesy of Ametek/Thermox Company.

quate sample flow. Sample enters the $\frac{1}{4}$-in. inlet tubing, is measured by the flowmeter, goes up the ceramic inlet tube and down the annular space between the inlet tube and inside the electrochemical cell, and finally exits the $\frac{1}{4}$-in. outlet tube. Air aspirators are sometimes used on the sample outlet lines to assist sample flow.

The cell output is temperature dependent, so a thermostatically controlled heater is provided. Zirconium oxide probes are sometimes combined with catalytic combustion sensors. The catalytic combustion sensor consists of a measuring and reference thermistor or filament. The measuring element is coated with a catalytic surface that burns any fuel that comes in contact with it while the reference element compensates for sample temperature and thermal conductivity changes. The elements are part of a bridge circuit, and as fuel is burned, the temperature of the measuring element increases, increasing the resistance and producing a bridge output potential proportional to the combustibles concentration, typically in the range of 0 to 5%. Both elements are heated and thermostatically controlled. Flow past the elements is balanced, and in some cases the heater and thermostat of the oxygen cell may be used for the combustion sensor as well.

Figure 6.3 shows a unique *in situ* combination oxygen–combustibles sensor that operates on the zirconium oxide cell principle. In the current

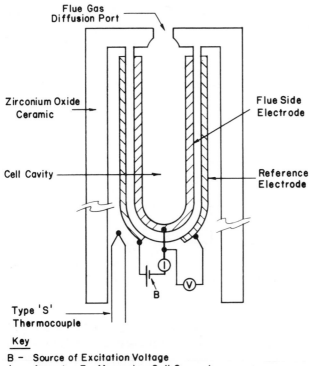

Flue Gas
Diffusion Port

Zirconium Oxide
Ceramic

Flue Side
Electrode

Cell Cavity

Reference
Electrode

Type 'S'
Thermocouple

Key
B - Source of Excitation Voltage
I - Ammeter For Measuring Cell Current
V - High Impedance Voltmeter

Figure 6.3. High-temperature combination oxygen–combustibles sensor. Courtesy of Westinghouse Electric Corporation.

mode, the cell is a cavity where the oxygen partial pressure of the diffused flue gas is reduced to near zero by the oxygen pumping action created by an applied excitation voltage. As oxygen diffuses into the cavity and is pumped out, positive current flows in the ammeter circuit that is proportional to the oxygen concentration in the flue gas.

When excess combustibles are present in the flue gas, the voltage produced by the cell exceeds the excitation voltage and the oxygen flow and current are reversed. Oxygen then flows from the reference side into the cell cavity and combines with the combustibles present. In this case, oxygen passes through the porous platinum electrodes and continuously purges the flue gas side electrode with protective oxygen even in the presence of strong reducing agents. Figure 6.4 shows typical sensor output characteristics as a function of oxygen or combustibles concentration. The continuous 4- to 20-mA output signal facilitates closed-loop control.

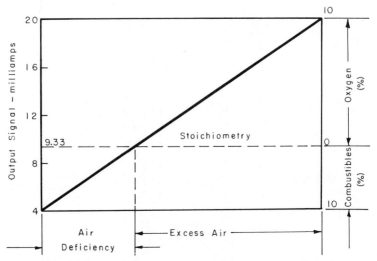

Figure 6.4. Milliampere output of combination oxygen–combustibles sensor. Courtesy of Westinghouse Electric Corporation.

The sensor can also act as a secondary standard when it is operated in the voltage mode with manual temperature compensation. This mode is used for calibration and verification of proper operation of the current mode. The cell output voltage is proportional to absolute temperature and the natural logarithm of the ratio of the partial pressure of oxygen in the reference air divided by the partial pressure of the oxygen in the flue gas as given in equation (6.2):

$$E = KT \ln \frac{P_1}{P_2} \qquad (6.2)$$

Where T is the absolute temperature in degrees kelvin, K is the proportionality constant, P_1 is the partial pressure of oxygen in reference air, and P_2 is the partial pressure of oxygen in the flue gas. In the voltage mode the excitation voltage source is not used.

With another model of the zirconium oxide oxygen analyzer, a stainless steel support tube with shield and flange is permanently mounted to the furnace wall or duct. Then the sensor unit and transmitter are inserted into the support tube and bolted in place (see Figure 6.5 for details).

The sensor unit consists of a 20-μm ceramic filter, insulated and temperature controlled nickel measuring chamber. Two zirconium oxide microcells are used to generate a voltage proportional to the difference in the

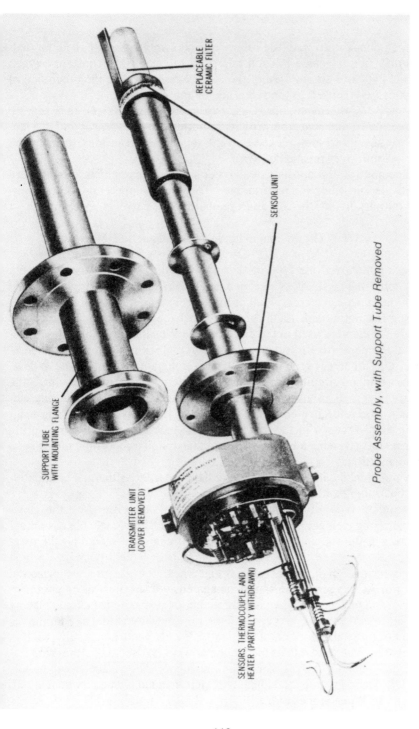

REPLACEABLE CERAMIC FILTER

SENSOR UNIT

SUPPORT TUBE WITH MOUNTING FLANGE

TRANSMITTER UNIT (COVER REMOVED)

SENSORS THERMOCOUPLE AND HEATER (PARTIALLY WITHDRAWN)

Probe Assembly, with Support Tube Removed

Figure 6.5. *In situ* oxygen probe with stack support tube and mounting flange. Courtesy of Leeds & Northrup Company.

oxygen concentration of the reference and sample gases. Palladium–palladium oxide is sealed in each microcell to establish a constant oxygen partial pressure on one side of the electrochemical cell; the other side of each cell is exposed to the reference or sample gas.

A type K thermocouple is used to keep measuring cell temperature above 700°C, and a weatherproof transmitter converts sensor and thermocouple signals to multiplexed current signals and transmits them to the control unit over unshielded wire.

The control unit consists of a power supply, preamplifier, temperature controller, and function programmer. The microcomputer provides a voltage output, digital display, and optional output currents and alarms.

6.1.2. Thermomagnetic and Paramagnetic Detectors

Figure 6.6 shows one type of thermomagnetic oxygen sensor assembly that consists of a stainless steel cell block with an internal cavity equipped with thermistor detectors. Half of the thermistor detectors are placed in a strong magnetic field. Each thermistor pair consists of a measuring and reference element used in the circuitry of a Wheatstone bridge. A proportional temperature controller holds the cell block at a constant temperature, and a current regulator provides a constant current to the Wheatstone bridge circuit. Sample and reference gas diffuses into the dead-ended cell cavity, where oxygen is attracted to the measuring elements. The reference elements are located just outside the strong magnetic field. Oxygen in the sample or reference gas cools the measuring thermistors, causing a bridge offset voltage that is proportional to oxygen concentration.

Similar cells without the magnetic field are used as thermal conductivity and combustible gas detector cells.

Another type of thermomagnetic oxygen analyzer uses the flow-through ring element[2] shown in Figure 6.7. As shown, sample gas enters the ring at the top and the oxygen in the stream is attracted by the magnetic field between the two poles. The resistor-wound filaments, which are two arms of a Wheatstone bridge, heat the gas in the horizontal section. As oxygen in the sample is heated, it loses most of its paramagnetism and new cool oxygen is attracted into the magnetic field, displacing the hot, demagnetized oxygen. As the hot oxygen-bearing stream moves to the right in the horizontal section of the element, a magnetic wind is generated. The flow rate of the magnetic wind is detected by the temperature difference of the resistor-wound elements. The magnetic wind cools the left-handed filament and heats the right-handed filament, which unbalances the Wheatstone bridge.

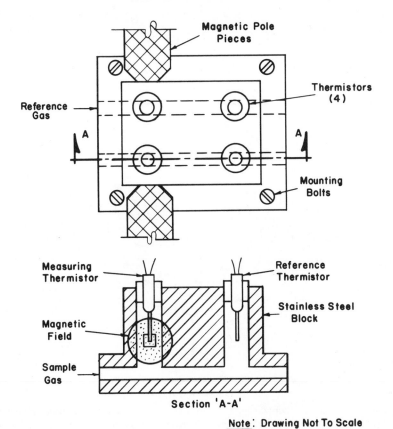

Figure 6.6. Thermomagnetic oxygen sensor consists of magnetic field impressed on thermoconductivity cell.

The sensor bridge, which is powered by a transistor-regulated power supply, is rebalanced and the amount of rebalancing required is proportional to the oxygen concentration. A temperature controller is used to provide a constant temperature environment for the bridge sensor.

Thermomagnetic oxygen analyzers are sensitive to pressure variations and diamagnetic gases such as hydrogen and helium.

Faraday demonstrated in 1851 that a hollow glass sphere filled with oxygen was attracted by a magnet. Later, it was also discovered that a hollow quartz sphere was slightly repelled by a magnet. In modern instruments the sensitivity of the paramagnetic–diamagnetic property is amplified by suspending dumbbell-shaped spheres in a symmetrical nonuniform magnetic field as shown in Figure 6.8. Since the hollow dumbbell spheres

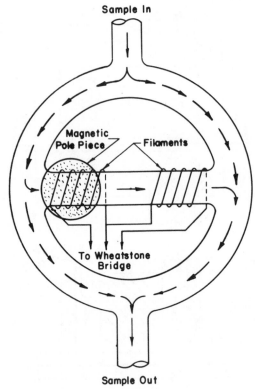

Figure 6.7. Ring-type thermomagnetic sensor creates a magnetic wind through the center of the detector. Courtesy of Mine Safety Appliance Company.

are slightly diamagnetic, they take up a zero position just to one side of the strongest part of the magnetic field. When sample gas enters the cell, the dumbbells are pushed further out of the magnetic field by the highly paramagnetic oxygen.

Oxygen is about 100 times stronger than diamagnetic gases such as methane, ethane, hydrogen, helium, carbon monoxide, and carbon dioxide. Therefore, equal or smaller quantities of these gases cause errors of less than 1%. However, other paramagnetic gases such as nitric oxide, nitrogen dioxide, and chlorine dioxide are strong interferences and will create substantial errors if present in more than trace amounts. The paramagnetic oxygen analyzer is not susceptible to other physical properties of gases.

In practice, the hollow-sphere dumbbell is held in place by a tough but slender rare metal suspension. The zero position of the dumbbell is

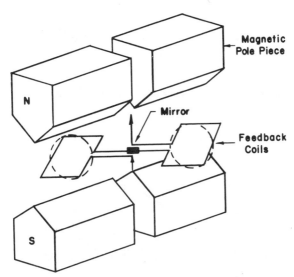

Figure 6.8. Evacuated dumbbell is the heart of the paramagnetic oxygen sensor. Courtesy of Taylor Instrument Company.

sensed by twin photocells that measure the light reflected from a small mirror on the center of the dumbbell. The output of the photocells is amplified and then fed back to a coil wound on the dumbbell so that the torque due to oxygen in the sample is balanced by a restoring torque created by the feedback current. The result of the restoring torque and suspension system is a heavily damped null-balance system. Figure 6.9 illustrates this principle of operation.

Paramagnetic oxygen analyzers whose detectors do not come in contact with the sample stream can be used with corrosive gas mixtures. One design of this type from West Germany is shown in Figure 6.10. This design also has the advantage that it is not affected by thermal conductivity.

The sensor shown in Figure 6.10 operates on the principle that a differential pressure is generated when two gases with different oxygen concentrations are mixed in a magnetic field.[3] Reference gas enters the sensor at the top and mixes with the sample gas along the sides of the sample gas chamber near the bottom. The fluctuating magnetic field generates a pulsating differential pressure at the pole piece that is equal and opposite to the differential pressure signal at the detector location. When the magnetic field increases the back pressure on the right side, a pulsating balancing flow is measured by the detector and converted into an electrical signal that is then amplified and conditioned by the electronics unit.

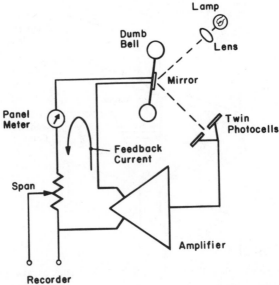

Figure 6.9. Simplified diagram of paramagnetic oxygen sensor. Courtesy of Taylor Instrument Company.

6.1.3. Calibration Techniques

Zirconium oxide stack gas analyzers have a calibration gas line for zeroing and spanning with cylinder gases. Instruments are typically zeroed with oil-free, instrument-grade nitrogen and spanned with various blends of oxygen with nitrogen gas mixtures. Cell output is typically 0 mV with air on both sides of the zirconium oxide element and about 120 mV when the calibrating gas contains 0.1% oxygen. A 2% oxygen in nitrogen standard would produce a 50-mV output. Analyzers of this type are generally calibrated for the range of 0 to 25% or 0.1 to 21% oxygen. A gas flow rate of about 2 liters/min is required. Some manufacturers offer a remote calibration panel with solenoid controlled valves for reference air and calibration gas, flowmeters, and light-emitting diodes (LEDs) to indicate when the reference or calibration gas flow is on or off.

Thermomagnetic and paramagnetic oxygen analyzers have a selectable range switch and can be calibrated for the full 0 to 100% oxygen range if desired. Some sample conditioning such as that shown in Figure 6.11 is generally required in order to present a clean, dry sample to the analyzer. One of two possible sample streams is transported through the demister–filter, secondary filter, and flowmeter into the analyzer. Part of the sample stream may be continuously bypassed to improve the system response

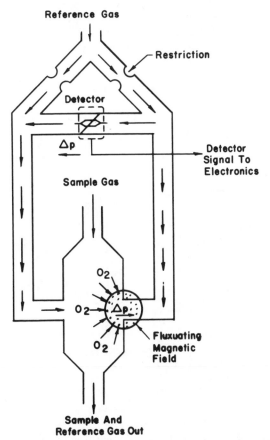

Figure 6.10. Microflow oxygen sensor detects small pulsating pressure changes. Courtesy of Siemens, Inc.

time, and the analyzer vent can be returned to the process if required. A small nitrogen purge is applied through a restricting orifice to the instrument case, and a differential pressure switch is used to warn of "loss of purge" and interlock the 110-V AC power when the instrument is used in a hazardous environment. Three-way manual selection valves are used to select zero or calibration gas on "Cal." or sample gas streams.

6.1.4. Design Features and Applications

Zirconium oxide stack gas oxygen analyzers are available with log scales of 21 to 0.1% oxygen or 0.1 to 21% oxygen and linear scales of 0 to 25% or

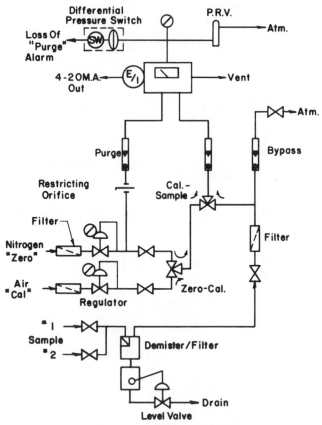

Figure 6.11. Typical sample system for paramagnetic-type oxygen analyzers. Courtesy of Taylor Instrument Company.

0 to 100% oxygen depending on manufacturer. Response time is typically 3 s for a 90% change in reading with an unfiltered probe or 13 s for a 63% change in reading when a 5- to 20-μm filter is used. Accuracies range from ±1% to ±5% of reading depending on instrument range and readout, with a repeatability of ±0.2% typical. Probes operate at 650 to 815°C, and cell warm-up time varies from 20 to 30 min. Reference and calibration gas flows of 4 scfh are typical. A wide range of current and voltage outputs is available.

Thermomagnetic and paramagnetic oxygen analyzers have 3 to 5 selectable ranges covering the 0 to 100% oxygen scale. Clean dry air is used as a reference gas. Accuracy is ±1 to ±2% of full scale. Sample flows of 1 to 2 liters/min are typical. Most of these analyzers operate at or near atmospheric pressure. Response times range from 1 to 7 s for a 90% change in reading. Warm-up times range from 10 min to 12 h depending on manufacturer. A wide range of current and output voltages is also available with these analyzers.

Applications

Zirconium oxide–type oxygen analyzers are widely used in combustion air control systems. The probes are either mounted directly in the flue gas stream, or stack gas is aspirated to the cell for analysis. High-temperature ceramic probes are capable of continuous operation at 1525°C. Figure 6.12 shows the distribution of products from the combustion of various fuels. An attempt should be made to optimize combustion efficiency by using the minimum amount of air required for complete combustion while fully utilizing the heat generated from the amount of fuel burned. In the case where natural gas and fuel oils are burned, a combination of oxygen and combustible analyzers are frequently installed to maintain carbon monoxide at very low levels, as shown in Figure 6.12, and about 10% excess air by volume would be needed, which would result in 2% excess oxygen by volume in the flue gas.

One strategy for controlling the combustion process consists of using oxygen control for gross tuning and carbon monoxide control for fine tuning in order to operate at maximum efficiency.[4] Carbon monoxide is formed only in the flame envelope and provides a direct measure of combustion efficiency. The inlet vanes of the combustion air fan would be modulated by the carbon monoxide analyzer as long as the flue gas oxygen content stays within preset high and low limits. When oxygen exceeds either limit, the control system switches to the oxygen analyzer. A dirty burner would cause the carbon monoxide content of the flue gas to increase without affecting its oxygen concentration. In this event the carbon monoxide analyzer might attempt to add air, but the oxygen controller would modify this action at the preset upper oxygen limit.

Thermomagnetic and paramagnetic oxygen analyzers also can be used for stack gas monitoring and combustion efficiency control. These analyzers are also used when oxygen levels in process streams are the controlled variable. In hydrocarbon processing air leaks can cause explosive gas mixtures. Oxygen analyzers are also used to monitor oxygen deficiency or enrichment in nonventilated or confined areas. Automobile exhaust gases are often monitored to determine whether carburetion is correct. Oxygen analyzers are frequently used to monitor inert gas purges to assure that oxygen will not degrade or contaminate storage tank contents or allow the formation of explosive or flammable gas mixtures. Airflow is carefully

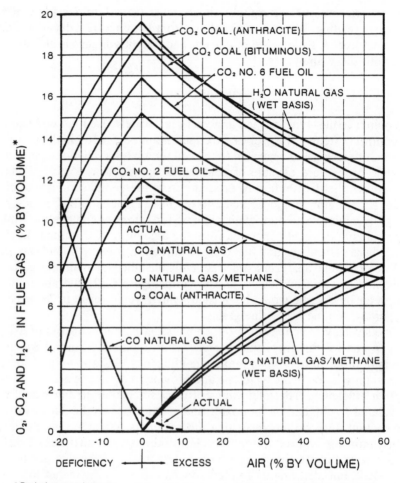

* Dry basis except as indicated.

Figure 6.12. The products of combustion as a function of air excess or deficiency.

controlled by oxygen analyzer input in catalyst regeneration systems when this air is introduced to the reactor bed to promote combustion in the presence of the hot catalyst.

6.1.5. General Comments

Microprocessors have been used with zirconium oxide oxygen analyzers to utilize multiple-probe arrangements to automate air calibration cycles and to provide probe or electronic diagnostics. The microprocessor can be programmed to produce a display of "9999" for an open thermocouple or "8888" for a sensor failure or to automatically switch the display to the cell temperature if the heater fails. These failures do not upset the combustion control as the last oxygen analysis is held in memory and alarms are activated to alert the operating technician of an analyzer problem.

6.2. LIQUID ANALYZERS

Dissolved oxygen analyzers are widely used to measure the dissolved oxygen content in sewage treatment plants, aeration ponds, waste treatment plants, boiler feedwater systems, and natural lakes and streams.[5]

6.2.1. Galvanic and Polarographic Dissolved Oxygen Analyzers

Galvanic and polarographic analyzers are similar with the exception that less noble metals are used for the anodes of galvanic instruments and the resulting electrochemical reactions are different depending on the anode material and electrolyte fill solution.

Table 6.1 summarizes some typical galvanic and polarographic reactions. It should be noted that the cathodic reaction is the same in all cases. The Clark polarographic equation [equation (6.5)] is sometimes listed as a galvanic reaction also.

Table 6.1. Summary of Galvanic and Polarographic Reactions

Galvanic		
Silver cathode	$O_2 + 2H_2O + 4e^- \rightarrow 4OH^-$	(6.3)
Zinc anode	$2Zn \rightarrow 2Zn^{2+} + 4e^-$	
Gold cathode	$O_2 + 2H_2O + 4e^- \rightarrow 4OH^-$	(6.4)
Copper anode	$2Cu \rightarrow 2Cu^{2+} + 4e^-$	
Polarographic		
Gold cathode	$O_2 + 2H_2O + 4e^- \rightarrow 4OH^-$	(6.5)
Lead anode	$2Pb \rightarrow 2Pb^{2+} + 4e^-$	
Gold cathode	$O_2 + 2H_2O + 4e^- \rightarrow 4OH^-$	(6.6)
Silver anode	$4Ag + 4Cl^- \rightarrow 4AgCl + 4e^-$	
Three-Electrode System		
Cathode	$O_2 + 2H_2O + 4e^- \rightarrow 4OH^-$	(6.7)
Anode	$4OH^- \rightarrow O_2 + 2H_2O + 4e^-$	

The galvanic dissolved oxygen analyzer shown in Figure 6.13 is designed to detect 0 to 100 parts per billion (ppb) of dissolved oxygen in a high purity water stream with a constant flow of 50 liters/h.[6] The sample flow is set by the flow regulator and monitored by the flowmeter before entering the sample conditioning tank. The sample conditioning or dosing tank is filled with low-solubility reagent crystals to keep solution conductivity above 2 μS and help keep the cell electrodes clean. When the calibrate cell and current generator are not being used to calibrate the instrument, the calibrate cell is used as a conductivity cell to warn of low-conductivity sample or the need to refill the dosing tank. The oxygen generator will not work properly if solution conductivity is too low. When the conductivity is less than 2 μS, the "Fill tank" lamp is illuminated. The thermistor in the sample stream is used to help provide automatic temperature compensation over the range 5 to 50°C.

The measuring cell consists of a silver cathode and zinc anode. A current proportional to the dissolved oxygen concentration is generated in accordance with the equations shown in equation (6.3).

The flow and sample conditioning system, thermistor, calibrate and measuring cells, lamp, and preamplifier are mounted in a rugged cabinet near the process, and the rest of the instrument components can be located in another cabinet within 500 ft.

An improved galvanic probe design is shown in Figure 6.14 that uses a copper anode.[7] Copper is an easy material to machine and forms a soluble product with the potassium hydroxide electrolyte. Other anode material and electrolyte combinations can cause insoluble precipitates to form at the anode, which can shorten cell life by plugging the membrane material. Likewise, oxide films that can form on other cathode materials and interfere with the electrochemical reaction do not form on the gold cathode. Teflon® is used as a membrane material because it resists fouling.

Other improvements for this probe design include rugged polyvinyl chloride (PVC) construction, a large electrolyte reservoir for long life, a stable polarization voltage-generated current relationship, good linearity, and uniform sensitivity. The probe reacts with oxygen in accordance with the equations shown in equation (6.4). Probes of this design have been in continuous service for over 1 yr.

The cell responds to molecular oxygen and generates a current that varies linearly with the dissolved oxygen content of the sample. When oxygen is not present, the gold measuring electrode is negatively charged because of the relative positions of the metals in the electromotive series as shown in Table 6.2. Because of this polarization and resultant abundance of electrons on the gold cathode no current flows in the external

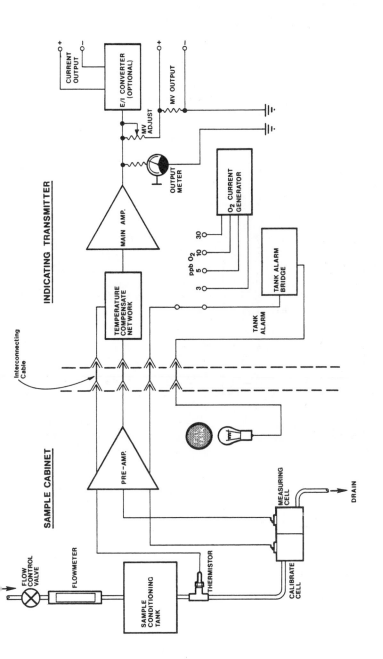

Figure 6.13. Parts per billion of dissolved oxygen analyzer features O_2 current generator for ease of calibration. Courtesy of Milton Roy Company.

131

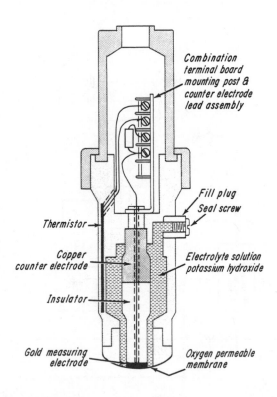

Figure 6.14. An improved galvanic dissolved oxygen probe with long-life electrodes. Courtesy of Fischer & Porter Company.

measuring circuit. Oxygen diffuses through the membrane and reacts with the gold cathode, leaving it partially depolarized so that some electrons can flow to it. At the same time electrons also flow from the copper anode, which oxidizes it to form cupric ions. The cupric ions form a hydroxyl complex with the electrolyte and stay in solution to help keep the cell in active working condition. The depolarizing current produced by the cell reactions flows through the load resistor to the measuring circuit in the transmitter. A thermistor near the outer surface of the probe constantly senses the solution temperature and sends its signal to the transmitter.

Probes of this design are generally inserted directly into the process stream. A minimum flow of about 0.2 ft/s is required for good response.

The unique three electrode system shown in Figure 6.15 features a

Table 6.2. The Electromotive Series[a]

Potassium	Nickel
Sodium	Tin
Barium	Lead
Strontium	Hydrogen
Calcium	Copper
Magnesium	Arsenic
Aluminum	Bismuth
Manganese	Antimony
Zinc	Mercury
Chromium	Silver
Cadmium	Palladium
Iron	Platinum
Cobalt	Gold

[a] Potassium most active; gold least active.

probe whose electrochemical reactions are completely reversible as given in equations (6.7).

The probe consists of a permanently encapsulated three-electrode arrangement. The cathode and anode are active electrodes that are interspaced on a supporting substrate and covered with the electrolyte. The permanent electrolyte is surrounded by a permeable membrane that is sealed at the rear with an expansion chamber to compensate for pressure changes. The third or reference electrode is in the center of the substrate and is also in contact with the electrolyte. Oxygen reduction and generation occur at the active electrodes, and the thermistor compensates for the oxygen solubility–temperature relationship. The reference electrode maintains the correct electrochemical potential.

When the probe is immersed in the sample stream, oxygen diffuses through the membrane and is reduced at the cathode. An equal amount of oxygen is simultaneously generated at the anode. Diffusion continues only until the oxygen partial pressure on both sides of the membrane is in equilibrium. The current necessary to maintain this equilibrium is directly proportional to the oxygen concentration. The current signal is amplified and displayed on the front panel meter directly in ppm of oxygen. Temperature of the sample stream can also be read directly on the front panel meter by turning the mode switch to "Temp."

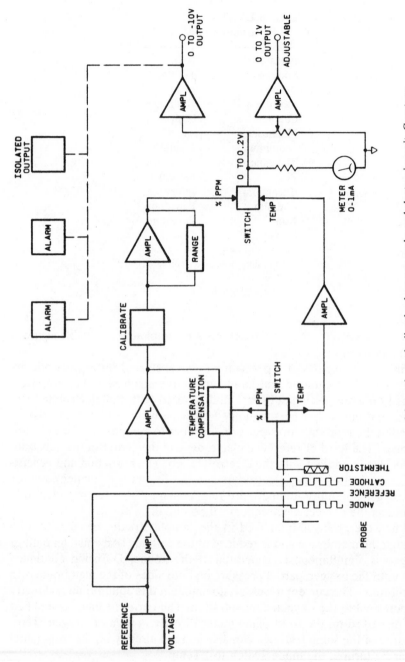

Figure 6.15. Block diagram of three-electrode dissolved oxygen probe and electronics unit. Courtesy of Leeds & Northrup Company.

134

Advantages of this type of probe are:

1. No membrane replacement required.
2. No consumption or contamination of the electrolyte.
3. No plating or etching of the electrodes.
4. No fouling precipitates formed.
5. No probe drying required for air calibration.

6.2.2. Calibration Techniques

The ppb galvanic dissolved oxygen analyzer, described in Section 6.2.1, is manufactured complete with its own oxygen generator, which works well as long as solution conductivity is maintained at 2 μS or higher. Another method for checking other galvanic and polarographic instruments is to measure the probe output when it is alternately placed in a solution saturated with sodium sulfite and a solution saturated with air at 25°C. The sodium sulfite removes essentially all the oxygen from the water sample, and the air saturated samples may be checked against standard air saturation tables.

All galvanic and polarographic oxygen analyzers can measure either gaseous or dissolved oxygen. In many cases an "air check" switch position and meter calibration mark are provided so that the dissolved oxygen probe can be removed from water and checked in air. Some systems require that the oxygen-permeable membrane of the probe be dried for a satisfactory reading.

6.2.3. Design Features and Applications

Membrane-type galvanic and polarographic oxygen analyzers are used to measure both dissolved oxygen and sample temperature.[8] Oxygen ranges of 0 to 20 ppm or mg/liter are common, with some manufacturers offering a 0- to 100-ppm dissolved oxygen range. Normal process temperature range is 0 to 50°C, with a few manufacturers offering a 0 to 100°C range. Accuracy and stability of ±1% of full scale is typical. The flow requirements for these analyzers have been greatly reduced in recent years and now are in the range of 0 to 0.2 fps. Many analyzers also have an "air calibrate" mode. Multiple output voltages and currents are available.

Applications

Galvanic and polarographic dissolved oxygen analyzers are used to monitor pollution in rivers, lakes, streams, and industrial plant effluents. Membrane-type dissolved oxygen probes do not require a sample system if they are submersed in a sample stream meeting the flow, pressure, and temperature requirements of the probe. Most probes of this design can be attached to a 1-in. galvanized or plastic pipe and mounted to a metal handrail or grating.[9,10]

Monitoring and improving the efficiency of aeration systems is a popular application for these systems. The average oxygen concentration of the aeration facility (3 to 5 points) is compared to a setpoint or desired value, allowing the oxygen controller to vary this concentration by controlling effluent flow, aeration motor speed, or lift blower vane position.[11] The mode of control depends on the process used and the size of the plant.

Other dissolved oxygen analyzer applications include:

1. Control of activated sludge basins and municipal water treatment plants where correct oxygen levels promote proper bacterial growth.
2. Monitoring beverage and food processing plants where oxygen can degrade food quality.
3. Monitoring fermentation processes where oxygen is required for proper bacteria growth.

6.2.4. General Comments

Present-day dissolved oxygen meters are reliable and relatively maintenance free as a result of improved sensor designs, electrolytes, and materials of construction.[12] Drift and stability problems have been greatly reduced through the use of non-precipitate-forming electrolytes and non-fouling membranes. Multiple dissolved oxygen analyzer installations will likely incorporate the use of microprocessors in the near future to improve the operation of aeration facilities and provide probe and analyzer diagnostics.

CHAPTER

7

PHOTOMETRIC ANALYZERS

This chapter is divided into four sections dealing with infrared (IR) and near-IR analyzers, ultraviolet (UV) analyzers, colorimeters, refractometers, and light-scattering photometers. All these analyzers operate in the electromagnetic spectrum shown in Figure 7.1. It is interesting that all the colors we see with the human eye fall within a relatively narrow band of the electromagnetic spectrum. Depending on the UV source and detector, UV analyzers may also be used at visible wavelengths, and these analyzers are thus sometimes referred to as UV–VIS analyzers. Colorimeters, refractometers, and light-scattering photometers also usually operate in the visible band.

The section dealing with IR and near-IR analyzers will also include information of IR moisture analyzers and CO_2 and pyrolysis-type analyzers for total carbon (TC) and total organic carbon (TOC) analyses.

This chapter does not deal with γ rays, x rays, or radio waves because their applications in continuous on-line analyzer systems are limited. Some notable exceptions to this would be the use of γ waves and radio waves in moisture analyses and the use of x-ray attenuation in density–concentration analyses.

UV and IR analyzers and optical densitometers follow Beer's law, which states that absorbance, or the logarithm of the light transmitted by a particular species, varies linearly with the concentration of that species. The term "electromagnetic energy" could be substituted for light, and optical density could also be substituted for concentration in Beer's law.

7.1. ULTRAVIOLET ANALYZERS

Photometric analyzers that operate in the UV band can operate equally as well in the visible band by changing their light source and optical filter arrangements. These instruments are sometimes instead catagorized by their functional groupings:

1. Specific gas, vapor, or liquid analyzer.

137

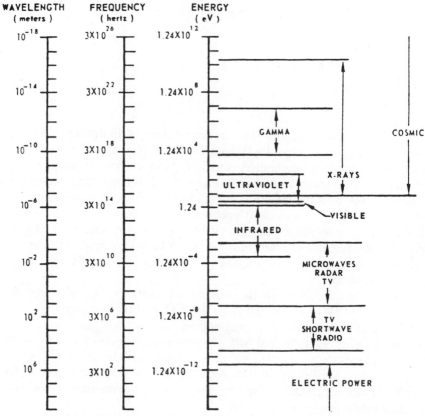

Figure 7.1. Wavelength, frequency, and energy relationship of the electromagnetic spectrum.

2. Film or coating thickness analyzers.
3. Process polarimeters.
4. Emission monitors.
5. Wastewater analyzers.
6. Optical densitometers or color analyzers.
7. Opacity, turbidity, haze, or suspended solids monitors.

The last two groups of analyzers are covered in Sections 7.3 and 7.4. Because of the large number of instruments involved in this chapter, an attempt is made to cover only the basic principles involved. No attempt is made to describe all the variations in design that are commercially available.

7.1.1. Theory of Operation

Figure 7.2 shows a typical split-beam UV analyzer that measures the difference in energy absorption by the sample at two different wavelengths in the UV region. Radiation from the UV source passes through the sample cell and into the photometer housing, where a semitransparent mirror splits the energy into two beams. One beam passes through the measuring beam filter, which excludes all wavelengths except the measuring wavelength. The measuring wavelength is chosen such that the component of interest in the sample stream strongly absorbs UV energy at that wavelength.

The other beam passes through the reference beam filter, which transmits only the reference wavelength. The reference filter is selected so that the component of interest in the sample stream absorbs the UV energy weakly or not at all at that wavelength. The reference filter must also be chosen so that other components in the sample stream do not absorb energy at that wavelength. The beam splitter can direct 30, 50, 70, or 90% of the energy to the measuring circuit and the balance to the reference circuit. The beam splitter is usually chosen so that the energy in both beams is nearly balanced when the zeroing medium is present in the sample cell. This balancing compensates primarily for the difference in transmitted energy of the UV source at the two different wavelengths.

Changes in source energy and small amounts of accumulated dirt in the sample cell have little effect on instrument accuracy as long as the foreign matter attenuates the measuring and reference signals to nearly the same degree. Therefore, the variations in the concentration of the component of interest have the greatest effect on the measuring beam intensity and instrument output.

Some UV absorbing materials are listed in Table 7.1. In general it is possible to look for or control these materials in other nonabsorbing or transparent background process streams. Minimum full-scale ranges vary from about 1 ppm to 0.33% for vapors and 5×10^{-6} to 235×10^{-3} g/liter for liquids. The sensitivity depends on the absorption characteristics of the material at the measuring and reference wavelengths, filter transmission characteristics, characteristics of the cell window materials, and cell pathlength considerations. Air and water vapor are transparent at most UV measuring and reference wavelengths, but the vapor pressure and temperature of the sample also affect gas calibration ranges.

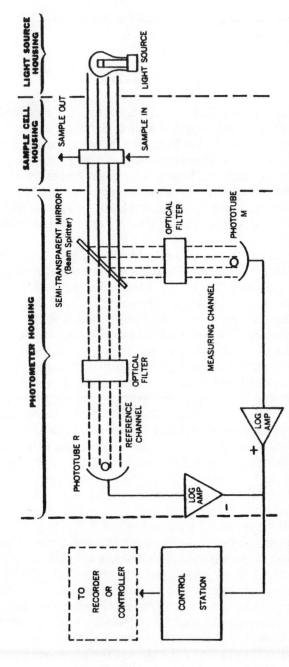

Figure 7.2. Split-beam UV analyzer with single light source and semitransparent beam splitter. Courtesy of DuPont Analytical Instruments.

140

Table 7.1. Partial Listing of UV Absorbing and Nonabsorbing Liquids, Vapors, and Gases

Absorbing

Acedic acid	Hydrogen sulfide
Acetone	Iodine
Ammonia	Mercury
Aniline	Methyl mercaptan
Anthracene	Naphthalene
Benzene	Nickel carbonyl
Bromine	Nitrobenzene
Carbon disulfide	Ozone
Carbon tetrachloride	Perchloroethane
Chlorine	Phenol
Chlorine dioxide	Phosgene
Chlorophenol (o, m, p)	Pyridine
Dioxane	Sodium sulfide
Ethylbenzene	Styrene
Ferric chloride	Sulfur
Fluorine	Sulfur dioxide
Furfural	Toluene
Hydrogen peroxide	Xylene (o, m, p)

Nonabsorbing

Acetylene	Hydrogen
Argon	Krypton
Butane	Methane
Butanol (i, n)	Methanol
Carbon dioxide	Neon
Carbon monoxide	Nitrogen
Ethane	Oxygen
Ethanol	Propane
Ethylene	Propylene
Ethylene glycol	Propanol (i, n)
Helium	Water
Hydrochloric acid	Xenon

Figure 7.3 shows a dual-beam instrument that is especially useful in chlorine purity measurements. A stream of high-purity (100%) chlorine is passed through the reference cell while the sample of unknown chlorine purity is passed through the sample cell. This arrangement enables a chlorine measuring range of 95 to 100% with an accuracy of ±0.1%. Changes in lamp intensity are automatically compensated for with the

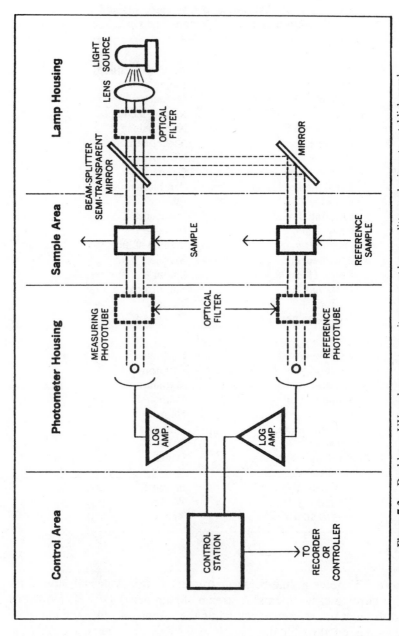

Figure 7.3. Dual-beam UV analyzer uses semitransparent beam splitter and mirror to establish parallel energy beams. Courtesy of DuPont Analytical Instruments.

142

dual-beam instrument, but dirt in either stream will cause an error in readings. Automatic zero standardization is usually required because of the high sensitivity and precision required. Three minutes each hour the sample flow is valved out and 100% chlorine is passed through both cells. During this period the fine zero control is automatically balanced so that the instrument output reads 100%. The zero standardization compensates for gradual long-term differences caused by ultrafine dirt, window plating or etching, aging of the detectors, and amplifier drift.

7.1.2. Calibration Techniques

UV, visible, and IR photometric analyzers operate according to the Lambert-Beer absorption law, which states that absorbance is:

$$A = abc = \ln \frac{I_{in}}{I_{out}} = \log \frac{1}{T} \qquad (7.1)$$

where A is absorbance, a is molar extinction coefficient or absorptivity, b is sample pathlength, c is sample concentration, I_{in} is energy intensity entering the sample, I_{out} is energy intensity leaving the sample, and T is percent of energy transmitted through the sample.

The absorption laws can be put into practical use in actual calibration work by taking into account the actual operating temperature and pressure of the sample gas or the molecular weight of the component of interest in liquid analyses. The six equations shown in Table 7.2 result when the proper correction factors are applied and consistent units are used. Only one variable can be unknown in each of the six equations.

The absorbance range (A) for full-scale instrument output is 0.25 to 4.0, with 0 to 1.0 A units a typical design range for new applications. The absorptivity (a) or percent transmission T can be obtained from experimental data, laboratory scanning instruments, or textbooks showing absorbance spectrum traces. The cell pathlength (l) equations are the most frequently used when instrument output range is changed and it is desirable to keep the absorbance the same. If the cell pathlength is to remain the same, a new absorbance range can be calculated so that the calibration point of a known absorber can be shifted accordingly.

Absorptivity varies with wavelength as shown in Table 7.3. In addition to hydrogen peroxide, phenol, n-heptyl-p-hydroxy benzoate, and sulfite ion UV absorption also varies with solution pH. There are ways to compensate for these variations as discussed in the applications section of this

144 PHOTOMETRIC ANALYZERS

Table 7.2. Absorbance Equations[a]

For Gases

$$A = \frac{(a)(l)(V\%)}{2450} \times \frac{psia}{14.7} \times \frac{298}{(°C + 273.15)} \qquad (7.2)$$

$$a = \frac{A \times 2450}{l \times (V\%)} \times \frac{14.7}{psia} \times \frac{(°C + 273.15)}{298} \qquad (7.3)$$

$$l = \frac{A \times 2450}{a \times (V\%)} \times \frac{14.7}{psia} \times \frac{(°C + 273.15)}{298} \qquad (7.4)$$

For Liquids

$$A = \frac{10(a) \times l \ (wt.\%)(sgu)}{MW} \qquad (7.5)$$

$$a = \frac{A \times MW}{10 \ (l)(wt.\%)(sgu)} \qquad (7.6)$$

$$l = \frac{A \times MW}{10(a)(wt.\%)(sgu)} \qquad (7.7)$$

[a] Symbols: A = absorbance units, a = absorptivity (liter/mole-cm), $V\%$ = volume or mole percent, wt.% = weight percent, psia = abolute pressure (gauge pressure + 14.7 lb), sgu = specific gravity units, MW = molecular weight, l = cell pathlength in centimeters.

chapter. In most of these systems dilute caustic or acid solution may be added if necessary to keep the absorption peak from shifting.

As a calibration example, let us assume that we are setting up a new analyzer system for chlorine gas analysis. Our desired instrument range is 0 to 5% chlorine by volume at 314.7 psia and 200°C. If we design the system for an absorbance range of 0 to 1.0, what would be the ideal pathlength? The measuring and reference frequencies must first be selected. Assume that a laboratory scan of chlorine-free process gas has shown that there are no other interferences at 313 or 405 nm. A book of UV scans should also be consulted for possible interferences from other gases in the process stream. Since chlorine strongly absorbs UV energy at 313 nm and only weakly absorbs it at 405 nm, the 313-nm wavelength is used as the measuring frequency and the 405-nm wavelength is used as the reference.

The next step is to calculate the absorptivity. In this case the absorptivity from Table 7.3 will be 54 minus 3.8 or 50.2 liters/mole-cm. The cell

Table 7.3.   Absorptivity versus Wavelength

Wavelength, nm[a]	Absorptivity
Chlorine Gas at 24°C	
313	54
334	62
365	27
405	3.8
546	0
Sodium Hypochlorite in Caustic	
254	42
265	109
280	210
313	166
334	52
365	4.5
405	0
546	0
Hydrogen Peroxide in Caustic (pH 10 or more)[b]	
254	234
265	122
280	75
313	0.4
334	0
365	0

[a] 1 nanometer (nm) = 1 millimicron (mμm).
[b] Hydrogen peroxide absorptivity varies with solution pH.

pathlength can then be calculated from equation (7.4) by entering the data:

$$l = \frac{1 \times 2450}{50.2 \times 5} \times \frac{14.7}{314.7} \times \frac{200 + 273.15}{298} \tag{7.8}$$

$$l = 9.76 \times 0.047 \times 1.59 \tag{7.9}$$

$$l = 0.73 \text{ cm or } 0.287 \text{ in.} \tag{7.10}$$

The instrument would then be set up with an inside window to window cell pathlength of 0.73 cm, a 313-nm measuring filter, a 405-nm reference filter, and a UV source lamp with a strong 313-nm emission band. To check the instrument, a 3% v/v chlorine in nitrogen-certified gas standard could be purchased and used as a span gas standard. A pure nitrogen gas cylinder could be used as a zeroing gas since it does not absorb UV energy. The zero and span or gain controls could then be adjusted so that the instrument output is zero on flowing nitrogen and 60% of full scale on the flowing 3% chlorine gas standard. Cylinder gas flow rates can be the same as sample gas flow rate, typically 2000 cm^3/min.

Since chlorine is an active gas and tends to change in time by reacting with the cylinder walls or absorbing into them, an attempt is usually made to locate a stable secondary standard. The secondary standard could be something as simple as a copper screen that is inserted into the measuring beam only or a piece of treated glass or plastic with different absorption characteristics at the measuring and reference wavelengths. Once a satisfactory standard or calibration filter is found, the use of gas cylinders is generally discontinued.

Some of the factors affecting instrument calibration in this example have been covered in detail. Most of the background and development work is done in advance by the instrument manufacturer since most small companies do not have the technical know-how, materials, or facilities of perform primary instrument calibrations or make significant calibration changes on installed instruments. However, a general knowledge of these basic principles should prove useful in discussing future problems with the manufacturer. The manufacturers of photometric equipment may also supply or recommend a secondary calibration standard for most applications.

7.1.3. Design Features and Applications

UV analyzers can be purchased in explosionproof or general-purpose housings and provided with long or short sample cells. The most flexible designs provide sample cell isolation so that the cells can be electrically heated or steam traced, whereas the photometer circuits are air or nitrogen purged and protected from sample system leaks or fume releases in the area. Most manufacturers also offer a number of UV light sources and power supplies so that the instrument can be used in a wide variety of applications. Operation to 1000 psig and 316°C is possible with the isolated sample cell design, and accuracy to ±2% with reproducibility to within ±1% is typical.

Figure 7.4 shows a special dual-beam analyzer for measuring hydrogen

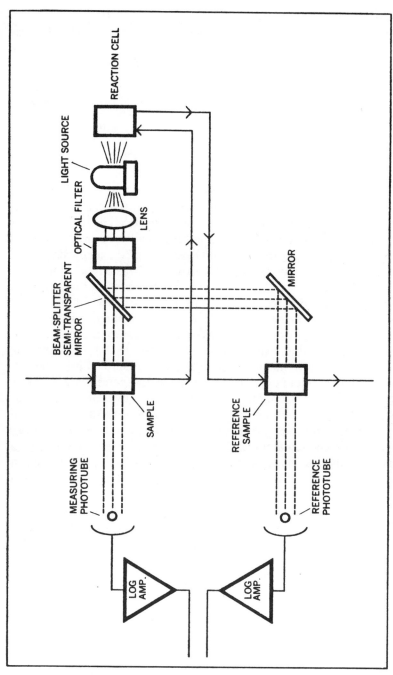

Figure 7.4. Special dual-beam analyzer uses UV reaction cell to react chlorine and hydrogen to form hydrogen chloride gas. Analyzer measures the difference in chlorine before and after reaction to determine the amount of hydrogen present. Courtesy of DuPont Analytical Instruments.

in chlorine. In this case the chlorine concentration is measured before and after a photochemical reaction. The sample flows through the sample cell, reaction cell, and reference sample cell. The hydrogen present combines with the chlorine in the presence of UV energy to form hydrogen chloride, which is transparent. The difference in chlorine concentration is a direct measure of chlorine reacted or an indirect measurement of hydrogen present. The photochemical reaction is given in equation (7.11):

$$H_2 + Cl_2 + UV \text{ energy} \rightarrow 2HCl \qquad (7.11)$$

The reaction cell is made of Pyrex® tubing so that the sample is exposed to the UV radiation. Low sample flow rates, typically 200 cm³/min, are used to assure that the reaction has sufficient time to come to completion. The residence time of the sample is shortened by bypassing most of the sample stream, typically 1800 cm³/min upstream of the analyzer.

The same analyzer can be used to measure hydrogen in a chlorine off-gas stream if there is a sufficient amount of chlorine present to react all the hydrogen present. To make sure that this is the case, the sample system shown in Figure 7.5 can be used. The sample-in is a potentially hydrogen bearing chlorine off-gas stream. A high-speed bypass loop allows most of the stream to bypass the instrument for improved speed of response. The bypass rotameter valve maintains enough back pressure on the sample to provide a 1000-cm³/min flow rate in the sample flow control branch of the sample system. At the same time a 1000-cm³/min flow rate of pure chlorine is added by the chlorine flow control branch of the sample system. The two streams are then mixed, and 1800 cm³/min is allowed to bypass the instrument. The remaining 200 cm³/min of sample gas passes through the analyzer, and cell pressure is maintained by the back-pressure regulator. Inlet pressures may be as high as 100 psi with back pressure regulated at 5 to 10 psi.

It is relatively important that the chlorine and sample flow rotameters be calibrated for actual gas density and balanced at a convenient midpoint flow rate. Changes in the ratio changes the dilution effect of the chlorine and adversely affects instrument calibration.

Table 7.4 lists typical applications for UV–VIS-type photometric analyzers and illustrates their versatility. Colorimeters are discussed in Section 7.3, and forward scatter turbidimeters are covered in Section 7.4.

7.1.4. General Comments

UV analyzers have been employed in industry since about 1940. They are by no means new instruments, and many of them still use the same basic

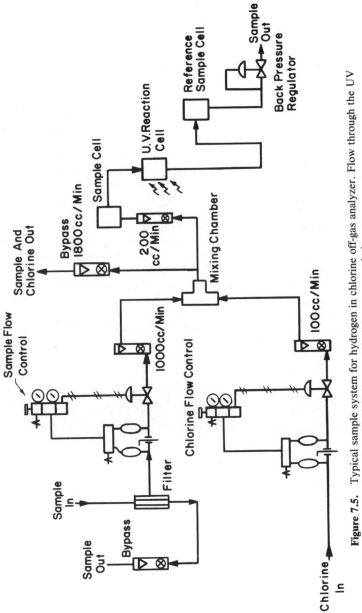

Figure 7.5. Typical sample system for hydrogen in chlorine off-gas analyzer. Flow through the UV reactor must be slow enough for the reaction to go to completion.

149

150   PHOTOMETRIC ANALYZERS

Table 7.4.   Applications

Application	Type of Analyzer
Trace chlorine	Split-beam
Chlorine purity	Dual-beam
Hydrogen analysis	Dual-beam with UV reactor
Color analysis	Split-beam
Film or coating thickness	Split-beam, no sample cell
Process polarimeter	Split-beam with polarizing filters
Opacity	Single-beam
Turbidity, forward scatter	Split-beam with condensing lens and mask
Turbidity, color correction	Dual-beam with filtered reference
Phenol in waste water	Split-beam
SO_2 in stack gas	Split-beam with sample system
NO_x emission monitor	Split-beam with sample system
H_2S in flare stream	Split-beam with sample system
H_2S/SO_2–sulfur recovery	Two split-beam analyzers with sample system

Source: Adapted from information supplied by Du Pont Analytical Instruments.

phototube measuring circuits as their original counterparts. The analyzers are extremely linear over a 10,000 to 1 optical transmittance range. They are also extremely rugged and are generally considered to be low-maintenance instruments. One manufacturer proudly proclaims that they have over 1000 units in service, some that have been in continuous operation for over 10 years.

UV analyzers are a reminder that there is still a very realistic need for a simple, single-channel, rugged analyzer in the process control industry. Even though microprocessor-controlled analyzers are playing an increasingly important role in this field, the dedicated single-channel UV analyzer has demonstrated more than 40 years of success.

7.2.   INFRARED ANALYZERS

Infrared analyzers are used to identify and determine the purity of a large number of chemical compounds in both liquid and gas forms. Specially adapted IR analyzers are used to measure component purity in high-viscosity or high-concentration liquids. This is done by a multiple reflection technique that will be described in detail later. Moisture or particle size of granulated or powdered solids can also be measured through the use of relatively elaborate optical systems. Some analyzers such as the total carbon analyzer oxidize and decompose small-volume liquid samples on a catalytic surface and measure the carbon dioxide produced with an IR analyzer to determine the total organic and inorganic carbon that was present in the sample. IR analyzers follow UV–VIS photometric analyzers in the analysis of components of interest having absorption bands in the electromagnetic wavelength energy spectrum.

7.2.1. Theory of Operation

When IR energy strikes nonsymmetrical molecules, it causes them to vibrate or rotate, resulting in IR energy absorption. These vibrational or rotational frequencies depend on the molecular structure of the chemical compound as shown in Figure 7.6, which illustrates chemical group vibration characteristics in the IR energy band of 2- to 15-μm wavelength. Figure 7.7 further illustrates the "fingerprint" region of a multicomponent organic stream. All the peaks in this IR scan are not identified, but a scan such as this can be used to determine which measuring and reference frequencies should be used for a dedicated field instrument. For example, if the peak at 5.7 μm were the component of interest, the IR instrument could be set up with a 5.7-μm measuring filter and 3.9-μm reference filter. Both filters pass only a narrow energy band (typically ± 0.03 μm) at the specified frequency and reject all other energy bands. The reference frequency is chosen such that none of the stream components absorb energy.

A partial list of IR absorbing compounds is given in Table 7.5. Uniform molecules such as oxygen, nitrogen, chlorine, and hydrogen as well as most rare earth gases do not absorb IR energy.

Figure 7.8 shows a simplified and more detailed view of the Luft-type nondispersive IR (NDIR) analyzer.[1] The instrument shown uses two nearly identical nichrome filaments as IR sources and projects the IR energy through parallel paths; one beam passes through the sample cell and the other beam passes through an inert gas-filled reference cell. The single detector cell is filled with a gas that absorbs energy at the measuring wavelength. The chopper motor alternately blocks the energy to the sample cell and comparison cell. The energy in both beams is balanced when the gas in both cells is transparent. When the gas entering the sample cell contains the component of interest, however, the energy in that path is

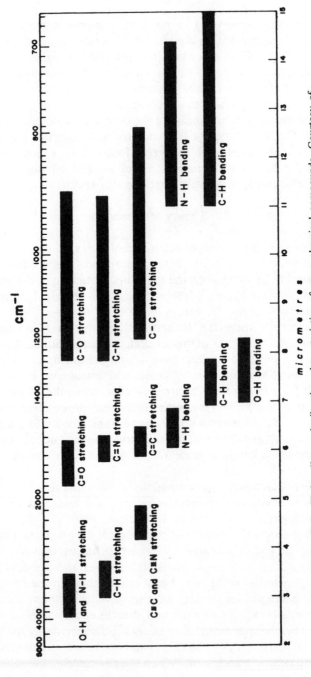

Figure 7.6. IR bonding and vibration characteristics of some chemical compounds. Courtesy of Foxboro Analytical.

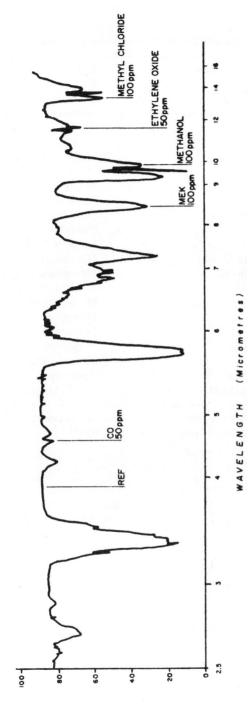

Figure 7.7. IR absorption characteristics of a multicomponent stream. Courtesy of Foxboro Analytical.

153

Table 7.5.   Partial List of IR Absorbing Compounds[a]

Carbon tetrachloride	Hydrogen cyanide
Chloroform	Hydrogen nitrate
Dichloromethane	Hydrogen sulfide
Dichloroethane (1,1 and 1,2)	Isobutane
Freon-13B	Methane
Freon-14	Methyl alcohol
Freon-C-318	Methyl azide
Butadiene (1,3)	Methyl chloride
Butane (n)	Methyl mercaptan
Carbon dioxide	Nitric oxide
Carbon monoxide	Nitroethane
Cyanogen	Nitrogen dioxide
Cyclopropane	Nitrogen pentoxide
Diazomethane	Nitromethane
Dimethyl amine	Nitropropane (1 and 2)
Dimethyl ether	Nitrosyl chloride
Dimethyl hydrazine (1,1 and 1,2)	Nitrous oxide
Ethane	Phosgene
Ethyl alcohol	Propane
Ethyl chloride	Propylene
Hydrazine	Trimethylamine
Hydrogen bromide	Trimethylhydrazine
Hydrogen chloride	Vinyl chloride

[a] Hundreds of IR absorbing compounds have been scanned with laboratory IR instruments and their spectra catalogued for IR applications work.

attenuated and less reaches the detector. This causes a radiation flicker that in turn causes the detector gas to alternately expand and contract. This movement varies the diaphragm of a condenser microphone, thus changing the capacity of the condenser and producing an electrical output signal proportional to the concentration of the gas of interest.

Another Luft-type IR instrument (not shown) uses a single IR filament source, concave focusing mirror for energy division, and dual-compartment gas-filled detector. The advantage of the single detector is that detector changes are self-compensating; the advantage of the single source is that source aging is self-compensating. The frequency response of both instruments is about the same. Luft-type instruments generally use gold-plated stainless steel sample cells and depend on both direct and reflected energy transmission for optimum performance with relatively long gas analyzer cell pathlengths.

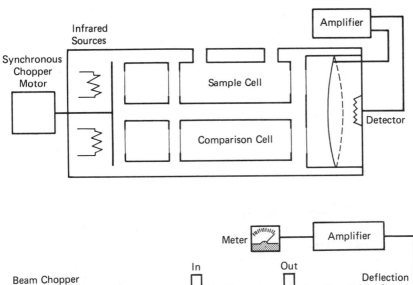

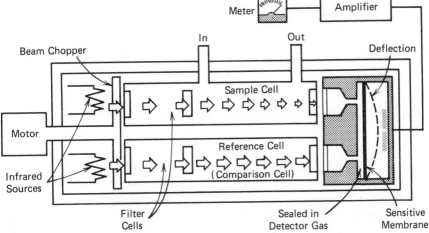

Figure 7.8. Simple and more detailed drawings of a Luft-type IR analyzer. Courtesy of Mine Safety Appliance Company.

The advantage of the Luft-type IR analyzer is its design features, which permit its use in the so-called fingerprint region of 2- to 15-μm wavelength and its excellent sensitivity to low gas concentrations. Disadvantages include the fact that optical alignment is critical and to some extent both electrical and mechanical balancing is required. Dirt or corrosion in the sample cell can also destroy the reflective properties of the sample cell walls, creating an imbalance that cannot be compensated for by the comparison cell. Since the Luft-type detector is a gas-filled capacitance microphone, it is also subject to vibration and occasional gas leaks.

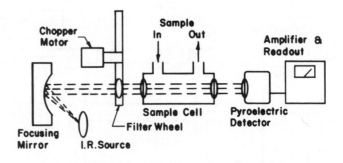

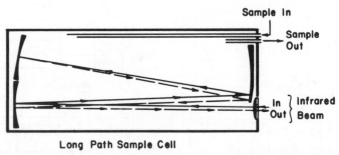

Long Path Sample Cell

Figure 7.9.   IR analyzer with multiple filter arrangement and long-path IR cell. Courtesy of
Foxboro Analytical.

Optical filters and solid-state detectors have eliminated some of these
problems with some corresponding sacrifice of sensitivity and bandwidth.
   Another IR analyzer that can be routinely applied in the 2.5- to 14.5-
μm range is shown in Figure. 7.9.[2] Energy from the IR source is focused
by a concave mirror, alternately passed through measuring and reference
filters, and then passed through the sample cell and finally detected by the
pyroelectric detector. The filter wavelengths are chosen so that the component of interest absorbs energy at the measuring wavelength and is
transparent at the reference wavelength. Dirt in the sample cell and
source and detector aging are automatically compensated for through the
use of a single source and detector and single alternating beam of energy
that passes through the sample cell. This design also permits the use of
multiple reflection and variable pathlength cells.[3] Calibrated dials are
used to adjust mirror position for pathlengths of 0.75 to 20.25m. These
long-pathlength cells result in improved detection sensitivity in gas applications; however, the gases must be noncorrosive with regard to the
mirror surfaces and mechanical positioning devices.
   The IR analyzer shown in Figure 7.10 evolved from the original Luft-

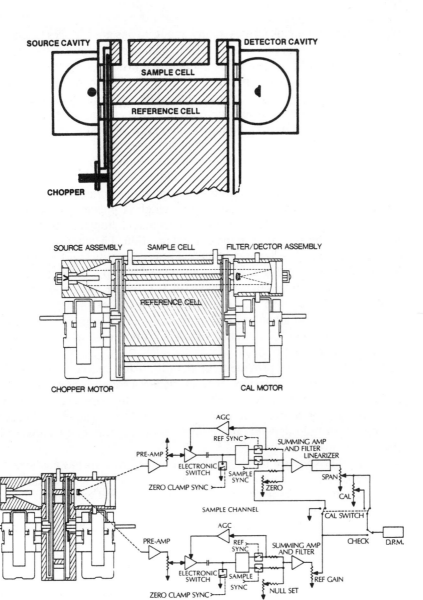

Figure 7.10. Top: sketch of tuned cavity type IR analyzer. Middle: IR analyzer for gas service showing chopper motor and calibration motor details. Bottom: IR analyzer for liquid service showing simplified schematic diagram. Courtesy of Anarad, Inc.

type design but incorporates many modern design features such as a single-cartridge source, optical filters, and a solid-state detector. The source consists of a nichrome filament housed inside a steel cartridge. The source cavity is shaped or tuned so that the IR energy is collimated through the sample and reference cells. The chopper motor alternately places the optical measuring and reference filters in the sample and reference beams. These narrow-bandpass filters are fabricated by vapor deposition techniques that apply as many as seven different coatings on a suitable substrate to achieve the desired transmission characteristics. A calibration motor inserts an optical calibration filter in the sample cell beam to obtain an upscale calibration point with a transparent medium in the sample cell. When the calibration switch is pushed, the calibration motor drives the filter flag against a mechanical stop that automatically places the calibration filter in the correct position. The chopper motor and calibration motor are identical; the calibration motor can be substituted for the chopper motor in case of a unexpected chopper motor failure. The solid-state detector is lead selenide (PbSe), which has an IR detection range of 2 to 5-μm. The electronic circuitry provides the necessary synchronizing and timing pulses and amplification to determine the measuring signal : reference signal ratio, which is proportional to the component of interest's concentration.

The reference cell provides compensation for source aging in cases where the energy emission characteristics change significantly with source aging. However, this dual-beam system does not provide compensation for film buildup on the sapphire cell windows. The effect of such a film buildup would depend on the thickness of the film and its absorption characteristics.

Figure 7.11 shows a highly selective double-beam IR gas analyzer that operates on an alternating energy principle.[4] The IR source operates at 700°C and emits IR radiation that is divided into two beams. The measuring or analytical beam passes through the sample cell chamber and into the receiver cell. Some of the energy in the measuring beam is absorbed by the gas component to be measured. The reference beam passes through the reference cell chamber and into the receiver cell. The chambers in the receiving cell are filled with the gas to be measured, thus sensitizing the analyzer to one particular gas component.

Gas in the right side of the receiver cell is heated more than that in the left side as a result of absorption of radiant energy in the sample. This generates a pressure differential, causing a balancing flow in the connecting chamber. The chopper wheel between the IR source and the filter cell interrupts both beams simultaneously, resulting in a pulsating balancing flow. The microflow sensor located at the pneumatic gravity center of the

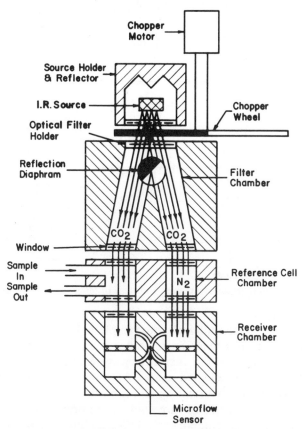

Figure 7.11. Sensitive IR detector using microflow sensor. Courtesy of Siemens, Inc.

receiver converts the flow pulses into electrical signals that are amplified into an analog output signal.

The microflow sensor consists of two nickel grid resistors sandwiched 0.15 mm apart. The two grid resistors are part of a Wheatstone bridge. A constant current through the bridge initially heats both resistors to about 100°C. The balancing flow through the microflow sensor cools the first grid resistor down while increasing the temperature of the second grid resistor. This temperature difference in grid resistors upsets the bridge balance, causing a flow proportional AC signal that is also proportional to the concentration of the component of interest.

The noncontacting IR analyzer, the optical measuring system of which is shown in Figure 7.12, was developed for making moisture in solids measurements. In this unit an extremely strong light beam is optically

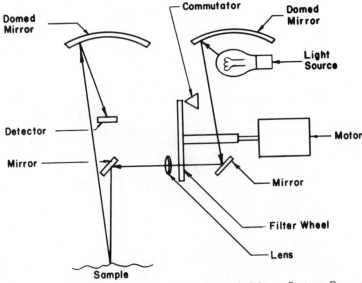

Figure 7.12.   Moisture in solids analyzer. Courtesy of Moisture Systems Corporation.

chopped and focused to illuminate the sample to be measured. The filter wheel contains the analytical and reference beam filters. Moisture in the sample absorbs energy at the analytical wavelength but not at the reference wavelength. The reflected energy from the sample illuminates the detector, providing an electrical signal proportional to the moisture content. Additional optical beams (not shown) are provided to compensate for variations in optical components. These additional channels provide greater accuracy and high-stability operation.

The signal from the detector is amplified and sent to a signal processor where it provides a true ratio of reference and reflected analytical energy levels. The signal is displayed digitally as percent moisture. This instrument is particularly useful in the pulp, paper, and fiberboard manufacturing industries for measuring surface moisture in applications where it is directly related to the total moisture content of the material being tested. The instrument can also be used to measure some film and coating thicknesses when these materials have absorption bands in the 2- to 5-μm energy band.

The multiple internal reflection (MIR) cell shown in Figure 7.13 has been developed for process applications involving high-viscosity or high-component-concentration liquids. This technique is also known as *attenuated total reflection* (ATR). As shown in Figure 7.13, the liquid sample is directed along an optical crystal through which an IR beam is reflected.

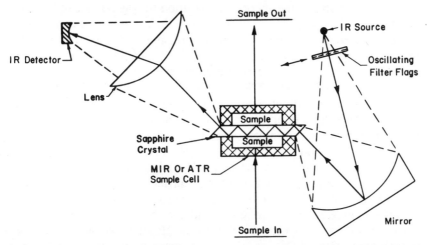

Figure 7.13. Drawing showing the principle of operation of the multiple internal reflection cell. Courtesy of Foxboro Analytical.

Oscillating filter flags establish alternating measuring and reference wavelength energy pulses. As the beam is reflected by the crystal, it is also attenuated by absorption in the sample. The amount of absorption depends on the measuring beam wavelength, the angle of incidence, and the relative indices of refraction. When a sapphire crystal is used, analyzer response is limited to the 2- to 6-μm wavelength range. The apparent beam penetration depth is less than half the measuring wavelength. This technique produces the same results as would be expected with a transmission cell having an extremely short pathlength. However, the MIR cell is easy to disassemble and clean and is not subject to most clogging problems. It is also relatively insensitive to particulate matter in the sample stream. Sample temperature must be carefully regulated so that it will not cause a measurable change in the index of refraction. In some cases an inert cabinet purge may also be required to prevent interference from vapors in the atmosphere surrounding the optics.

In the mid-1970's rapid and accurate methods for continuous monitoring of organic pollutants in wastewater were developed. Analyzers measuring total carbon (TC) and total organic carbon (TOC) replaced earlier and more complex total oxygen demand (TOD) and chemical oxygen demand (COD) analyzers in many applications. These new analyzer systems also reduced analysis time from 3 h in some cases to about 5 min.[5] Both UV oxidation and advanced high-temperature combustion systems are now manufactured in rugged industrial packaged systems.

Figure 7.14 shows the theory of operation for the advanced high-temperature combustion TC and TOC analyzers. Total carbon is the sum of inorganic carbon (IC) and TOC; therefore, the TOC analyzer must first remove any inorganic carbon before making its analysis. Operation of the TC analyzer is as follows: a continuously flowing sample is delivered to the analyzer at a flow rate of 50 to 1000 cm³/min at 1 to 5 psig. The sample flows through an overflow reservoir to a drain to shorten sample residence time. At preset time intervals the instrument aspirates a discrete sample from the reservoir and injects it into the 900°C reaction chamber. The sample combusts on a palladium catalyst surface, and the carbon contained in the sample is converted to CO_2, which is one product of complete combustion. Purified nitrogen carrier gas flows continuously through the reaction chamber and sweeps the combustion products through a gas scrubber to remove soluble and corrosive impurities. The carrier gas then transports the CO_2 to the IR analyzer, which measures the concentration of CO_2 at the 3.0-μm wavelength. The signal is then amplified, displayed, and transmitted to other optional devices as required by the user.

The TOC analyzer has the same sample supply requirements as the TC

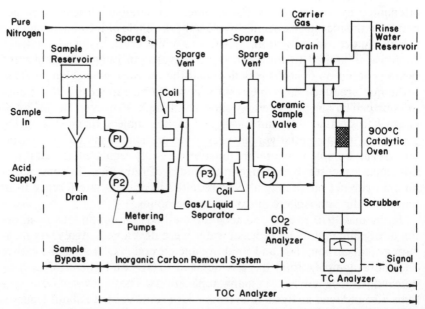

Figure 7.14.   Sample system and configuration of high-temperature TC and TOC analyzers. Courtesy of Ionics, Inc.

analyzer. In this case the sample is continuously pumped from the overflow reservoir to the two-stage inorganic carbon removal system where the sample and 2 N sulfuric or hydrochloric acid are continuously metered together to obtain a pH of 2 or less. Inorganic carbon in the sample is converted to CO_2 in suspension that is removed by nitrogen sparging in a mixing coil. The sample is then pumped to the second stage of the carbonate removal system, where additional nitrogen sparging occurs, resulting in a 99.5% IC removal efficiency. The sample is then pumped to the ceramic sample injection valve and analyzed by the TC analyzer, which is an integral part of the TOC analyzer.

7.2.2. Calibration Techniques

The absorption equations given in Section 7.1.2 also apply to IR and near-IR (NIR) analyzers. Laboratory-type scanning IR instruments are often used to obtain absorptivity values at various measuring wavelengths. In some cases the user may wish to purchase calibration gas cylinders with certified concentrations of the gas component. Nitrogen is frequently used as a zeroing gas since it is inert and transparent to IR. However, in most well defined applications manufacturers now supply coated optical filters that provide an upscale calibration point simulating a known component concentration. Pyrex®, quartz, ordinary window glass, and many plastics also have absorption bands that can be used to simulate some component concentrations. In single-beam systems the calibration filter chosen must absorb more energy at the measuring wavelength than at the reference wavelength.

Calibration filters are used in a number of ways; they can be manually inserted in front of the source or detector or inserted automatically by a filter flag–solenoid arrangement. In cases where they are manually inserted, definite storage areas should be provided so that the filters are not misplaced or lost. Coated optical calibration filters should be checked periodically by laboratory scanning techniques to assure that their transmission characteristics have not changed because of fume releases in the area or chemical or physical damage.

Care must also be exercised as to how the analyzer system is calibrated. Is the sample cell in the beam during calibration? Is the analyzer calibrated in air? What effect will dirty sample cell windows have on instrument calibration or routine operation? Is the manufacturer's built-in calibration check an actual check on measuring wavelength absorption or merely an electronic amplifier response test? These are some of the questions that should be of concern to the person who is responsible for instrument calibrations. In large plants periodic laboratory samples may also be taken and analyzed to confirm analyzer calibrations.

7.2.3. Design Features and Applications

Modern dual-beam Luft-type NDIR analyzers have a broad spectral range, reach 90% of final readings in 3 to 5 s, have noise levels of less than 1% of full scale, and have zero and span drifts of less than 1% of full scale in 24 h. Repeatability is within ±1%, and linearity is within 5 to 10%. An optional linearization circuit can be provided to correct the calibration curve response to within ±1% of straight-line response. Calibration curves are provided with each instrument and calibration is generally accomplished by using known gas standards. A precision resistor in the IR source circuit provides a convenient upscale span check. A standard adjustable millivolt output of 0 to 10 mV DC up to 0 to 100 mV DC is provided. Optional current outputs of 0 to 1, 0 to 5, 1 to 5, 0 to 20, 4 to 20, 0 to 50, or 10 to 50 mA DC are available.

Luft-type analyzer sample cells are aluminum blocks with gold-plated stainless steel tubular inserts. Optional nickel and Monel® cell and cells with pathlengths of up to 20 in. are available. Present-day instruments are constructed and mounted so that they are not affected by normal plant vibrations.

Dual-wavelength single-beam analyzers feature a broad-spectral range, solid-state pyroelectric detector, built-in self-diagnostics, and cell pathlengths of up to 50 cm. Drift is less than ±1% of full scale in 24 h, and noise and repeatability are within ±1% of full scale. Solid-state circuitry is provided, and optional features include automatic zero, dual-range operation, linearizer, and special cells for gas and liquid streams. Outputs of 0 to 2 V DC, 0 to 10 V DC, or 4 to 20 mA DC are available.

Dual-beam analyzers with tuned cavities and solid-state PbSe detectors have a spectral range of 2 to 5 μm and accuracy, resolution, and drift specifications similar to the instruments previously described. Other features include internal calibration, linearized outputs, digital displays, and high-pressure (1000-psi) stainless steel sample cells for liquids. A microprocessor can provide compensation for barometric pressure and ambient temperature changes along with signal conditioning, automatic zero and span calibration, and alarm programming. The unit can also communicate with a host computer.

The IR instrument with the microflow IR sensor has a span drift of less than 1% per week. Response time is 2 to 9 seconds for a 90% reading with sample flow rates of 1000 cm³/min. Sample cells can be heated to 110°C and supplied with tantalum linings for handling corrosive gas streams.

This unit also features complete separation of the sample cell and electronics, which is an advantage when handling corrosive gases. The IR sensor is also constructed so that it is not exposed to the sample gases and automatic zero and span gas calibration can be provided.

The noncontacting moisture in solids NIR analyzer has an adjustable moisture range from 0 to 0.2% to 0 to 90%. Accuracy is 0.5% of full scale, and repeatability is within 0.2% of full scale. Stability is such that calibration is required about twice a year. Sample to sensor distance can range from 3 to 12 in. Outputs of 0 to 100 mV, 0 to 10 V, 1 to 5 mA, 4 to 20 mA, or 10 to 50 mA DC are standard. Explosionproof enclosures can be provided if required.

Applications

Single- and dual-beam IR analyzers with broad spectral ranges are more versatile than narrow-band instruments because they can be used to analyze chemical compounds that have major absorption bands at longer-wavelength regions. Some analyses cannot be made in the NIR range because of interferences from other components in that relatively narrow band. Some chemicals with major absorption bands above 5 μm are carbon tetrachloride, chloroform, dichloromethane, hydrazine, methane, nitric oxide, nitrogen pentoxide, and phosgene. Some of these compounds have minor peaks below 5 μm but greater sensitivities can be achieved by monitoring them at higher wavelengths. Figure 7.15 shows the IR absorption characteristics of carbon tetrachloride, hydrazine, and nitric oxide. Carbon tetrachloride has absorption peaks at 6.5, 8, 10, and 13 μm; a major hydrazine peak occurs at 10.5 μm, and nitric oxide has an absorption peak at 5.3 μm. Broad spectral range instruments with suitable filters also work equally as well in the NIR range in addition to being able to detect and measure components having absorption bands at longer wavelengths.

In some cases the range of NIR analyzers has been extended to 5 or 6 μm, where they are now frequently used to measure moisture, anhydrous HCl, carbon monoxide, carbon dioxide, sulfur dioxide, ammonia, methane, and many other hydrocarbons. They can also be used to measure the concentration of any componenent having one or more absorption bands in the NIR region. NIR analyzers are finding increased use in moisture measuring applications and automotive emissions work.

NIR analyzers can be used to measure the amount of hydrogen in chlorine product or chlorine off-gas streams in chlorine manufacturing facilities where hydrogen and chlorine are separated. As mentioned ear-

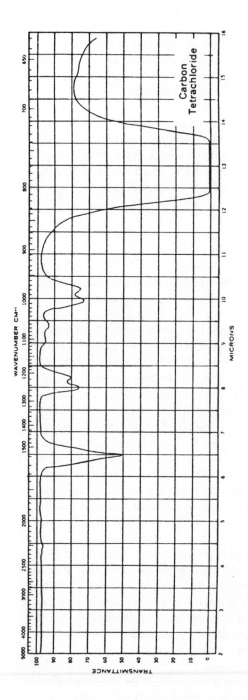

Carbon Tetrachloride

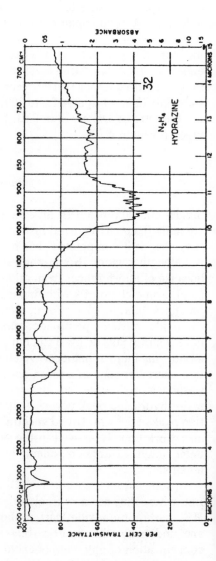

32

N_2H_4
HYDRAZINE

166

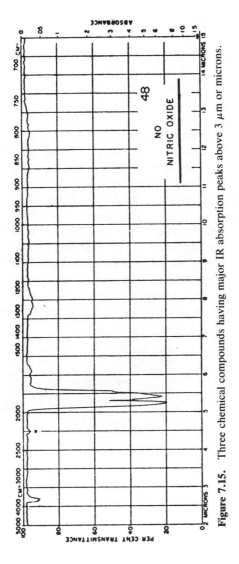

Figure 7.15. Three chemical compounds having major IR absorption peaks above 3 μm or microns.

167

lier, a UV reactor can be used to react hydrogen and chlorine in a small sample stream to form HCl in accordance with equation (7.11), where

$$H_2 + Cl_2 + UV \text{ energy} \rightarrow 2HCl \qquad (7.11)$$

In this case the measuring range of the NIR analyzer is 0 to 10% v/v HCl, which is equivalent to 0 to 5% v/v H_2.

Drying ovens are frequently used to dry paints, plastics, inks, and adhesives used in various manufacturing processes.[6] The drier outlet air often becomes laden with solvents that can pose a toxic, flammability, or air-pollution problem. At the beginning of the solvent recovery system the concentration of solvents can be monitored and controlled at safe levels by adding clean, dry dilution air as needed to maintain a safe control point. Figure 7.16 shows this control with a typical solvent recovery system where the solvent-laden air is absorbed on activated carbon beds, which also remove most toxic components. Eventually the carbon beds become saturated, and solvent breakthrough begins to occur. Samples of stripped air from the carbon beds are monitored by the IR analyzer, which can detect minute quantities of solvent and automatically switch a loaded carbon bed from the normal operating mode to a steam purge mode, which removes the solvents and transports them to a condenser where a solvent–water mixture is formed. The solvent mixture is then distilled to

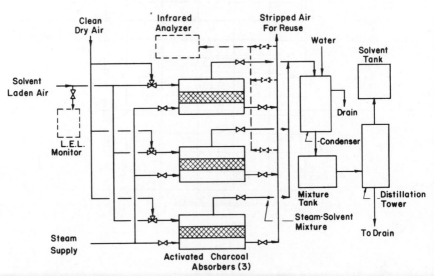

Figure 7.16. IR analyzer monitors effectiveness of solvent recovery system. Courtesy of Mine Safety Appliance Co.

recover the solvent while water is simultaneously drained from the system. After purging with steam the carbon bed is dried with clean, dry air and automatically switched back to the normal operating mode.

7.2.4. General Comments

MIR technology has found increasing use in many process applications where measurements are now made by process gas chromatographs (GCs). Some of these applications, previously regarded to be too difficult for IR analyzers, include the measurement of sugar and dissolved CO_2 in the beverage industry, the measurement of undesirable transisomers produced when hydrogenating vegetable oils, the determination of hydroxyl number in viscous polymers, the measurement of isocyanate groups in polyurethane production, and the measurement of ethanol in gasohol production. Because of the noncontacting, nondestructive nature of IR, it can also be used in the food processing industries. Another advantage of the MIR technique is that the output signal is continuous, resulting in a much lower overall analytical deadtime, which is very important in closed-loop process control applications.

To be truly competitive with gas chromatography, IR analyzer systems must be capable of multistream, multicomponent analyses in the broadband fingerprint region of the IR spectrum. IR analyzer systems are approaching this ideal through the use of programmed solenoid operated stream selector valves, multifilter filter wheel assemblies, and microprocessor storage of multistream component calibration data. It seems likely that continued progress will be made in these fields over the next several years.

7.3. COLORIMETERS

A number of simple one-dimensional color scales have been developed for grading liquid products in many process industries.[7] Two main types of colorimeter are discussed in this section; the first type uses UV–VIS analyzers with narrow- or broad-band filter response in the 400- to 700-nm visible band to make color comparisons with various color standards such as described in ASTM method D1500 for the ASTM color scale or ASTM method D156 for the Saybolt color scale. These analyzers are widely used in the petroleum refining business to measure the degree of refining and in the pharmaceutical industry to measure product purity or detect product aging.

The second type of colorimeter uses wet chemistry methods and various reagents that cause the process liquid to turn color or deepen in color when the component of interest is present. These analyzers are essentially chemical titrators with color-sensitive photodetectors. They can be used to measure many dissolved variables such as hardness, silica, or phosphates.

7.3.1. Theory of Operation

The split-beam colorimeter is identical to the UV–VIS analyzer shown in Figure 7.2. These colorimeters take advantage of the fact that simple one-dimensional color scales have been developed and accepted for grading liquid products in many process industries. One-dimensional color scales have resulted in continuous on-stream color monitoring to these standards. Colorimeters are now used to control intermediate process steps to assure the uniformity and high quality of final products. Analyzer readout is nearly linear with color intensity and relatively unaffected by bubbles, foreign matter, or light-intensity fluctuations. Analyzers that measure the color of refined products generally do not use reagents to change the color of the process liquids.

Figure 7.17 shows ASTM absorbance–wavelength curves. The dashed

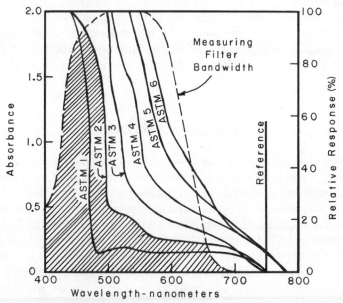

Figure 7.17. ASTM absorbance–wavelength curves showing measuring bandwidth of wide-band UV filter. Courtesy of DuPont Analytical Instruments.

line represents the sensitivity of the broad-band measuring filter and roughly approximates the response of the human eye or "standard observer" in the 400- to 700-nm band. The ASTM scale is used primarily for grading the color of lubricating oils in the petroleum refining industry. It is sensitive to colors ranging from a very dark red–brownish-black to near-water white. Analyzer response is nearly linear over the 0- to 6-ASTM color scale. The reference filter has a narrow bandwidth centered at 750 nm; at this frequency the ASTM colors have relatively little absorption compared to the measuring phototube–filter combination. The cross-hatched area represents a liquid having a color response equal to ASTM 2.

The Saybolt color scale is another widely accepted one-dimensional standard. Absorbance–wavelength curves are shown in Figure 7.18. The Saybolt scale is based on matching the color of a column of liquid sample with Saybolt color standards that are color-controlled glass disks. Saybolt color grades are typically measured by measuring the difference in absorbance at 436 and 546 nm. As shown, the bandwidth of both the measuring and reference filters are very narrow.

The absorbance–wavelength curves for two other color scales have not been shown; they are the Rosin color scale and the Alpha color scale.

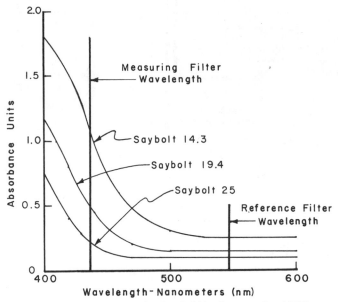

Figure 7.18. Saybolt color scale showing bandwidth of narrow band UV measuring and reference filters. Courtesy of DuPont Analytical Instruments.

Filters for the Rosin color scale are similar to filters for the ASTM color scale, and filters for the Alpha color scale are similar to filters for the Saybolt color scale. Rosin product also becomes lighter in color as it undergoes various refining operations. The Rosin scale covers colors from a dark reddish-orange to pale yellow. The Alpha (American Public Health Association) scale primarily grades the degree of yellowness in pale products.

The sample handling system for one type of wet chemistry colorimeter is shown in Figure 7.19. This colorimeter continuously samples liquid from a flowing process stream and adjusts the flow rate with a built-in needle valve and flowmeter. The sample flow chamber holds and continuously flushes a 60-ml sample. A float switch in the sample chamber detects a loss of process sample to automatically stop the analysis cycle. The float switch is bypassed when the sample is dumped into the reaction chamber. At the beginning of each 5-min analysis cycle the process liquid is valved into a reaction chamber and siphoned off to retain a 20-ml sample. Reagents are then sequentially injected into the sample by cam-driven metering syringes. The driving cams have a spiral shape so that the

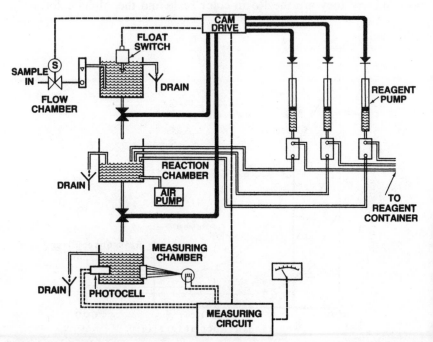

Figure 7.19. Wet chemistry colorimeter uses precision reagent pump syringes. Courtesy of Milton Roy Company.

syringes are slowly filled and then quickly emptied by spring force. Air is bubbled into the reaction chamber to assure mixing of the reagent and sample and then transferred to the measuring chamber. The color measuring system consists of measuring and reference photocells that are focused on the same spot of the lamp filament. The maximum volume of liquid in the measuring cell is 11 ml. A track-and-hold feature is provided, allowing the output of the instrument to be held at the last analytical value between readings; this provides a continuous output signal for process control.

Circuitry for the colorimeter is shown in Figure 7.20. The lamp control maintains constant brightness. The reference photocell detects any brightness changes and provides feedback for voltage control. The reference photocell compensates for lamp aging and uniform dirt films. A fullscale output is produced when the lamp circuit fails. The reference and sample photocells form part of an amplifier bridge circuit where the ratio of light intensity reaching each photocell is converted into a voltage. The signal is sampled for 5 s during each analysis cycle and stored on a capacitor. The output of the sample and hold circuit feeds a log amplifier that produces a linear output signal proportional to color intensity. For calibration, zero is provided by the log amplifier, and span is adjusted at the final amplifier. The alarm and current output modules are optional.

Figure 7.21 shows another version of the continuous, on-line colorimeter that uses precision, positive displacement pumps to transport and dispense the sample and reagents. Sample is taken from the process stream at any rate from 10 to 100 ml/min. A small sample is drawn from the sample stream every 30 s by the sample pump, and excess sample overflows to a drain. As the sample passes through the analyzer, reagents are introduced in precise volumes by reagent pumps. Chemical reactions and color development occur as the mixture flows through a carefully selected volume of sample tubing. The volume of the tubing is chosen so that the colorimetric reaction goes to completion prior to making the measurement.

7.3.2. Calibration Techniques

One or more optical filters may be used to calibrate linear-absorbance-type colorimeters. However, the Saybolt color grading system is based on matching the color of a column of liquid sample with color-controlled glass disks. After this has been done the instrument can be calibrated in Saybolt color units.

Wet chemistry type of colorimeters are precisely calibrated at the factory. However, premeasured dry chemical ampules are often used for

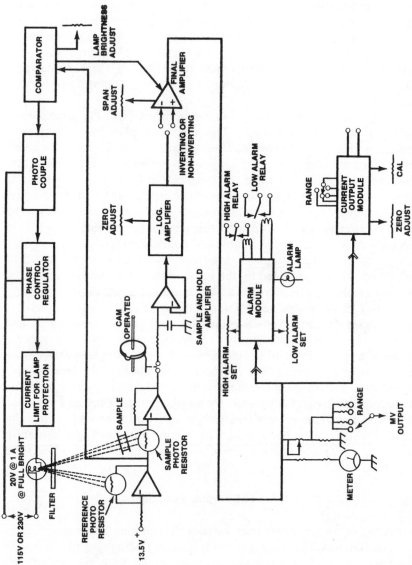

Figure 7.20. Block diagram of wet chemistry colorimeter. Courtesy of Milton Roy Company.

174

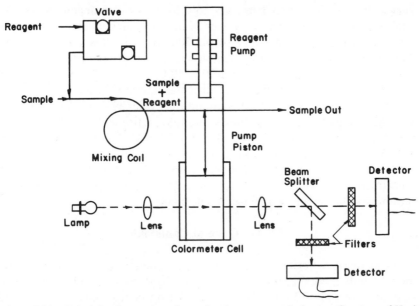

Figure 7.21. Principle of operation of wet chemistry pump colorimeter. Courtesy of Hach Company, Loveland, Colorado.

making secondary standards of known concentration so that the instrument range can be checked and adjusted if necessary. Instruction manuals also contain information on "standardizing" the instrument. A typical standardization for hardness might consist of:

1. Adding a known reagent to the measuring cell and observing the color produced.
2. Adjusted the zero control to produce an output indication of zero.
3. Adding a second known reagent to the measuring cell and observing a color change.
4. Adjusting the span of full scale control to produce a full-scale output.

7.3.3. Design Features and Applications

The design features of colorimeters that do not add reagents to the sample are identical to the features previously described for UV–VIS analyzers. The accuracy of the wet chemistry type of colorimeter varies from ±2% to ±5% depending on calibration range and analysis. Full-scale ranges vary from 0 to less than 1 mg/liter to 0 to 50 mg/liter depending on the analysis. Drift is typically ±1% over 24 h. The time between consecutive readings or the cycle time varies from 30 s to 5 min depending on manufacturer and design. Required sample flows vary from 10 to 150 ml/min with a sample temperature range of 0 to 50°C and a sample pressure of 10 psig. One manufacturer offers a colorimeter operable at sample pressures of up to 150 psig. Reagent consumption is typically 4 liters/month, and shelf life of most reagents is about 6 months. Instrument outputs of 0 to 10 mV, 0 to 100 mV, 0 to 1 V, 0 to 10 v, 0 to 20 mA, 4 to 20 mA, or 10 to 50 mA DC are standard with some manufacturers and optional with others.

Table 7.6 lists a number of applications and ranges for wet chemistry colorimeters showing that most of these applications apply to water conditioning systems and water treatment plants.

7.3.4. General Comments

Wet chemistry colorimeters use moderate amounts of reagents (4 liters/ month), indicating that they cannot run for long, unattended periods. The cost of reagents and the effect of interfering compounds must also be considered when applying these analyzers. For example, phosphate measurements are affected by silica concentrations greater than 50 mg/liter and other ion concentrations greater than 1000 mg/liter; silica measurements are likewise affected by phosphate concentrations greater than 50 mg/liter. Significant improvements have been made to these analyzers in recent years; however, these improvements consist primarily of more reliable, lower-maintenance sample metering and valving arrangements.

7.4. REFRACTOMETERS AND LIGHT-SCATTERING PHOTOMETERS

Refractometers measure the concentration or density of solutions in process streams. In critical-angle refractometry the process fluid is scanned optically by a sensing head mounted on the process line and the solution density is directly related to the index of refraction of the solution. The refractive index is determined by sweeping a beam of light through a series of angles at the surface of the crystal–solution interface. At some point in the cycle the angle between incident light and the surface of the solution decreases to the point where the beam is reflected back into the optical head instead of being refracted. This point of change depends on the refractive index of the solution and is called the *critical angle*. A light sensor detects the critical angle, and the output of the detector is con-

Table 7.6. Colorimeter Ranges and Applications

Parameter	Ranges, mg/liter	Application
Alkalinity, phenolphthalein	0 to 50 as $CaCO_3$ 0 to 100 as $CaCO_3$	Demineralized influent
Alkalinity, total	0 to 50 as $CaCO_3$ 0 to 100 as $CaCO_3$ 0 to 500 as $CaCO_3$	Process Water, boiler water
Chelant (EDTA)	0 to 2 chelant 0 to 2 hardness	Boiler water
Chlorine dioxide	0 to 0.5 0 to 1	Potable water and wastewater
Chlorine, free residual	0 to 0.5 0 to 1 0 to 2 0 to 5	Potable water and wastewater
Chlorine, total residual	0 to 0.5 0 to 1 0 to 2	Potable water and wastewater
Chromate, hexavalent	0 to 0.2 0 to 1	Cooling towers wastewater
Hardness	0 to 0.5 as $CaCO_3$ 0 to 1 as $CaCO_3$ 0 to 2 as $CaCO_3$ 0 to 5 as $CaCO_3$ 0 to 20 as $CaCO_3$ 0 to 50 as $CacO_3$	Demineralizers, softeners, boilers
Manganese^{7+}	0 to 0.2	Potable water, boiler water
Molybdate	0 to 10	Cooling towers
Ozone	0 to 0.5 0 to 1	Potable water
Permanganate, potassium	0 to 5 0 to 10 0 to 20	Potable water, boiler water
Phosphate, ortho	0 to 3 0 to 10 0 to 30	Boilers, wastewater
Silica	0 to 25 0 to 50	Boiler water

Source: Courtesy of the Hach Company, Loveland, Colorado.

177

verted to indicate the refractive index or percent concentration of the unknown variable.

Light-scattering photometers or turbidimeters continuously monitor process liquids for solid particles or colloidal contaminants that absorb or scatter light in accordance with Beer's law. The resultant decrease in light energy intensity or the increase in scattered light intensity is a measure of the turbidity of the solution. Light-scattering photometers measure either the light scattered at right angles to a strong collimated beam, the light scattered at or near the surface of a liquid when a light beam is bounced off the surface at a shallow angle, or the light scattered in a forward direction through defocusing of a strong collimated light beam. Light scattering and defocusing increases as turbidity increases.

7.4.1. Theory of Operation

The critical-angle refractometer shown in Figure 7.22 can serve as an on-line monitor for a wide variety of process liquids. Light from a tungsten lamp is collimated to produce parallel light, and a helical scanner rotates at 60 rps to move the light beam across a focusing lens. The focusing lens focuses the light at the interface of the sapphire prism and the process liquid. The angle at which the light enters the process fluid varies with the scanner position. At high angles of incidence the light is refracted into the solution; at shallow angles of incidence the light is reflected back through the sapphire prism to the solar cell detector. The ratio of refracted to reflected light varies with the refractive index of the solution. Light and dark periods are detected by a photocell at a 60-cycle rate because the rotational speed of the scanner is 60 rps. The electronics unit compares the light and dark periods seen by the solar cell and conditions the signal to produce a readout as refractive index (RI), specific gravity, weight percent concentration, or degrees Brix, Baume, or Balling depending on the application.[8] The unit is temperature compensated because the refractive index of a solution varies with temperature.

Light-scattering photometers measure the amount of light scattered at right angles, from the surface of a liquid, or in a forward direction. Figure 7.23 shows a low-range turbidimeter that measures the amount of light scattered by suspended solids such as clay, mud, algae, rust, or bacteria. The turbidimeter shown uses a submerged photocell to measure the amount of light scattered at a 90° angle to an incident light beam. Manufacturers of this type of turbidimeter claim greater sensitivity, precision, and accuracy. This turbidimeter is also known as a *nephelometer,* and its units of measurement are known as *nephelometric turbidity units* (NTU). The nephelometer can measure trace amounts of turbidity down to 0.04

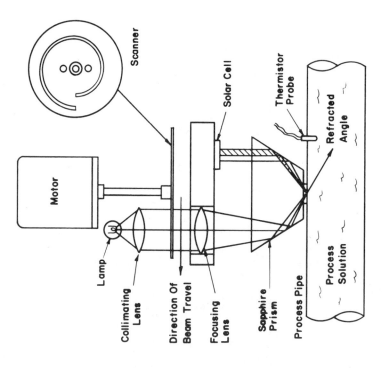

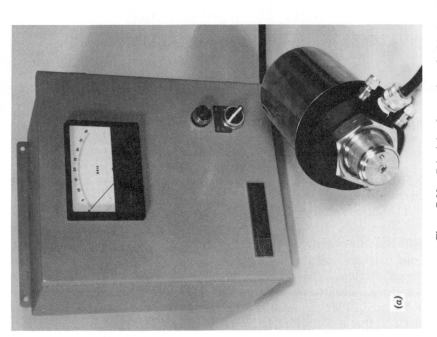

Figure 7.22. Principle of operation of critical-angle refractometer. Courtesy of The Electron Machine Corporation.

179

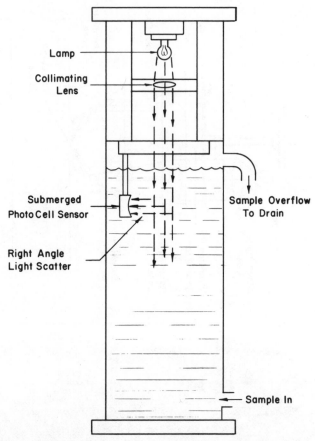

Lamp

Collimating Lens

Submerged PhotoCell Sensor

Sample Overflow To Drain

Right Angle Light Scatter

Sample In

Figure 7.23. Low-range–right-angle scatter turbidimeter. Courtesy of Hach Company, Loveland, Colorado.

NTU. The sample enters near the bottom and overflows above the submerged photocell.

The surface scatter turbidimeter shown in Figure 7.24 measures turbidity over a wide range. Very turbid samples can be measured because the light penetrates only a short distance beneath the surface of the liquid. Solid particles near the surface reflect and scatter light that is detected by the photocell mounted directly above the liquid surface. The turbidimeter is factory calibrated against formazin standards. The components shown in Figure 7.24 are mounted in a two-compartment box that is painted flat black so that it is unaffected by ambient light.

Figure 7.25 shows a typical forward-scatter turbidimeter used in appli-

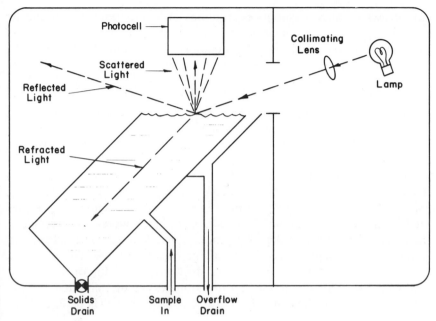

Figure 7.24. High-range–surface-scatter turbidimeter. Courtesy of Hach Company, Love-land, Colorado.

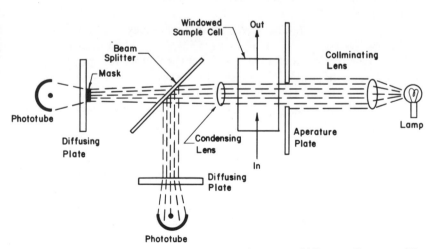

Figure 7.25. Principle of operation of the forward-scatter turbidimeter. Courtesy of Du-Pont Analytical Instruments.

181

cations where high sensitivity is required, color variations may occur, or use of a separate reference beam is not practical. The turbidimeter can be used on any sample that is not opaque. The light from the lamp is collimated and then passed through an aperature to further reduce uncollimated light. The light then passes through the cell to a condensing lens and beam splitter where part of the light is transmitted and part is reflected to the reference phototube. Light passing through the beam splitter is focused on a diffusion plate ahead of the measuring phototube. An opaque mask the size of the focused spot is placed on the measuring diffusion plate.

When the sample is not turbid, light passing through the condensing lens is focused on the diffusion mask and blocked from the measuring phototube. As the turbidity in the sample increases, defocusing occurs and a halo is formed around the mask. As turbidity increases, defocusing increases and more light strikes the measuring phototube. The reflected or reference beam slightly decreases with increasing turbidity. The reference beam compensates for instability and drift due to buildup or dirt on the cell windows. The forward-scatter turbidimeter can be calibrated in Jackson turbidity units (JTUs), which are equivalent to NTUs or directly in NTUs.

Figure 7.26 shows one way of decreasing turbidimeter response to color variations. With the use of broad-band light and appropriate filter–detectors, the turbidimeter can be made to be more sensitive to light in the NIR band from 750 to 1100 nm because color variation takes place mainly in the 400- to 750-nm visible wavelength band.

7.4.2. Calibration Techniques

Critical-angle refractometers are preset at the factory for specific applications. Zero, span, and AGC adjustments are provided for field calibration. The AGC adjustment is made by placing a midrange solution on a clean prism and adjusting the AGC control to obtain a predetermined DC voltage at specific test points in the circuit. The zero adjustment is made by placing a known low-range standard on the prism and adjusting the zero control until the meter reads correctly. Finally, a sample near the high end of the range is placed on the prism and the span control is adjusted to obtain the correct meter reading. The low- and high-range samples can be retested as necessary to obtain the best calibration possible.

A calibration kit is optional for the low-range 90° light-scattering turbidimeter. The kit consists of a portable calibration cylinder; hand-held digital titrator; and cartridges containing stable, concentrated formazin solution. Known formazin solutions can be made up and compared

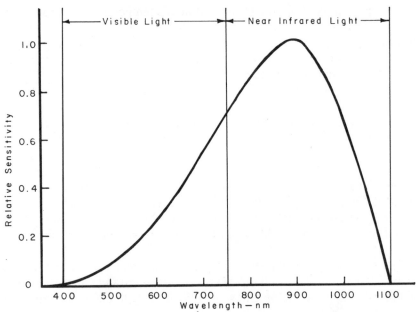

Figure 7.26. Visible and near-IR response of turbidimeter that minimizes its response to color. Courtesy of Siemens, Inc.

against a laboratory turbidimeter. Secondary reflectance rod standards made of frosted glass are also available. The rods are inserted through a self-sealing flexible septum and a specified NTU increase should be observed on the panel meter.

The surface scatter turbidimeter is factory calibrated against formazin standards; however, permanent glass turbidity reflectance plates can also be provided as a convenient secondary standard for routine calibration checks. The flat glass plates are placed on top of the sloped sample tube so that light is reflected off the glass plate. The surface of the plate is etched so that some of the light is scattered in the direction of the photocell. The etching on each surface of the glass plate is carefully controlled so that its scattering effect corresponds to a known NTU value marked on that surface.

7.4.3. Design Features and Applications

Critical-angle refractometers can be accurate and repeatable to within ±1% of span or ±0.0001 RI, whichever is greater. They can measure the refractive index of solutions at process temperatures of up to 350°F with process pressures of up to 300 psi. Automatic temperature compensation over a 40°F range is normally provided. Standard outputs are 0 to 100 mV or 0 to 1.0 V. Optional current outputs of 1 to 5, 4 to 20, and 10 to 50 mA DC are available. Other optional features include dual alarms, track-and-hold output, steam prism cleaner with automatic purge cycle, and loss of cooling water alarm. Cooling water is used to maintain the optical system at a temperature of 50 to 80°F with a cooling coil water flow of 1 liter/min.

The low-range nephelometer has full-scale ranges of 0 to 0.2, 0 to 1.0, 0 to 3.0, and 0 to 30 NTU. Accuracy and repeatability vary from about ±2% at the higher ranges to ±0.04 NTU at the most sensitive range. Voltage outputs of 0 to 10 mV, 0 to 100 mV, 0 to 1 V, and 0 to 10 V are standard; current outputs of 0 to 20 mA, 4 to 20 mA, and 10 to 50 mA DC are optional. Sample flow is typically 0.25 to 0.5 gpm. Other optional features include an alarm module with dual setpoints, remote indicator, floor stand mount, high pressure (≤150 psi) installation kit, and formazin calibration kit.

Full-scale ranges of the surface scatter turbidimeter are 0 to 1.0, 0 to 10, 0 to 100, 0 to 1000, and 0 to 5000 NTU. Accuracy and repeatability are within ±2% of scale. Sample flow is typically 0.25 to 0.5 gpm at ambient temperature and pressure. Voltage and current outputs are the same as in the low-range nephelometer. Other features include a linear broad range, noncontacting optics, nonblinding response even at high turbidities, and permanent secondary calibration plates.

The forward-scatter turbidimeter, also called a *suspended solids monitor,* has full-scale ranges of 0 to 10, 0 to 25, 0 to 100, and 0 to 250 ppm or 0 to 100, 0 to 250, 0 to 1000, and 0 to 2500 ppm of suspended solids. The monitor can also be calibrated in NTU's. Accuracy depends on good calibration techniques, and repeatability is within 1% of full scale. Voltage outputs of 0 to 100 mV DC and 0 to 10 V DC are standard, and current outputs are optional. The acceptable sample temperature range is 5 to 45°C, and sample flow rate can vary from 8 to 40 liters/min. Other optional features include sample conditioning equipment, digital panel meter, remote outputs and alarms, automatic zeroing, and remote range selection.

Applications

Table 7.7 lists the refractive index of various process solutions at 68°F. The critical-angle refractometer provides an accurate readout for binary mixtures in relatively dirty process streams. However, impurities with RI's near the component of interest can severely limit the usefulness of the critical-angle refractometer.

Critical-angle refractometers are used to:

Table 7.7.   Refractive Index of Binary Solutions

Acetic acid	1.3718	Formic acid	1.3714
Acetone	1.3588	Glycerol	1.4729
Acrylic acid	1.4224	Glycol	1.4318
Amyl acetate	1.4012	Heptane	1.3876
Benzene	1.5011	Hexane	1.3749
Butyl acetate	1.3951	Hexanol	1.4135
Butyl alcohol	1.3993	Hydrazine	1.4700
Butylene	1.3962	Hydrogen chloride	1.2560
Carbon disulfide	1.6295	Lead tetraethyl	1.5198
Carbon tetrachloride	1.4631	Menthol	1.4580
Chlorobenzene	1.5248	Methyl alcohol	1.3288
Chloroform	1.4464	Methylethyl ketone	1.3807
Cycloheptane	1.4440	Nitric acid	1.3970
Cyclohexane	1.4262	Nonane	1.4055
Cyclohexanone	1.4503	Octane	1.3975
Cyclopentane	1.4065	Pentane	1.3575
Decane	1.4120	Perchloroethylene	1.5053
Diethyl benzene	1.4955	Phenol	1.5425
Dimethyl benzene	1.4972	Propanol (n)	1.3851
Diethyl ether	1.3497	Propanol (iso)	1.3776
Ethyl acetate	1.3722	Styrene	1.5434
Ethyl alcohol	1.3624	Toluene	1.4969
Ethyl benzene	1.4952	Water	1.3330

Source: Courtesy of The Electron Machine Corporation.

1. Measure solution density or concentration, even in the presence of color, opacity, foam, turbulence, entrained air, suspended particles, or varying pressure.
2. Monitor black liquor soluble solids during recovery boiler operations in the manufacture of pulp and paper.
3. Measure and control the Brix content of citric juices, soft drinks, jams, jellies, and coffee concentrate.
4. Control the consistency of catsup, tomato sauce, and tomato paste.

Typical applications for turbidimeters in sewage treatment plants include:

1. Measurement of activated sludge content in activation tanks for automatic control of returned sludge volume.

2. Continuous monitoring of solid particles in the discharge of industrial or municipal purification plants.
3. Measurement of solid particle content of plant influents.
4. Control of sediment level in secondary sedimentation tanks.

Applications in monitoring potable water include:

1. Turbidity measurements in drinking water processing plants.
2. Monitoring of solid particles in water tanks, rivers, and lakes (1 NTU is the turbidity limit set by the U.S. government for drinking water standards).

Applications for various industrial and chemical processes include:

1. Control of solid particle concentration in filter water for the contro uniform purification and flotation.
2. Continuous monitoring of chemical reactions such as crystallizati fermentation, and precipitation.

Applications in the food and beverage industry include:

1. Monitoring and control of filtering and centrifugal processes in breweries.
2. Turbidity measurements in the production of fruit juices and sweetened beverages.

7.4.4. General Comments

Keeping analyzer optics clean and providing automatic temperature compensation are keys to obtaining accurate results with the use of refractometers; keeping the optics clean and providing compensation for color, lamp, and photocell aging are important factors for accurate monitoring with turbidimeters. The surface scatter turbidimeter has the advantage that it has no windows or sensors that are subject to collecting dirt but the system is limited to relatively low sample flow rates and must be used at atmospheric pressure.

CHAPTER
8
DIGITAL ANALYZERS

Since the advent of the digital computer and microprocessor, an increasing number of analyzer systems could be described as digital analyzers. We have already seen how microprocessors are being used to increase the accuracy and versatility of stack gas analyzer systems and provide multiple-probe monitoring in complex corrosion monitoring systems. This chapter describes in detail three analyzer systems currently using microprocessors to increase their versatility and flexibility. The chapter is divided into three subsections entitled "Automatic Titrators," "Digital Densitometers," and "Octane Analyzers." As is seen later, the computing units used with the automatic titrator and digital densitometer can be used to perform other functions as well.

Digital analyzers must be easy to communicate with; they should be easy to program, debug, and repair. Program sequences that are specific for the application of the digital analyzer are usually stored in PROM or EPROM memory. Much larger, more complex programs that can control analyzers and condition their raw data, perform calculations, and drive communications terminals usually require higher-density memory devices such as bubble memory or core. In very specific applications like the gasoline octane analyzer, programmed memory may not be accessible to the user.

8.1. AUTOMATIC TITRATORS

The automatic titrator described in this chapter[1] is a multifunction digital analytical instrument suitable for making up to three simultaneous measurements on process solutions containing inorganics, reactive organics, or dissolved metals. Some of the measurements that can be made are acid–base, Karl Fischer, or amperometric titrations; change of color end point or color absorption characteristics; and specific ion concentrations. Titrators of this nature are generally mounted in a field analyzer house where sample line runs can be kept short and where air, electric, water, and process drain or vent connections are readily available.

Automatic titrators are wet chemistry analytical instruments comparable in complexity to present-day process gas chromatographs. These units are expensive and cannot be run for long unattended periods, but their reliability has been increased to the point that the major maintenance effort is usually directed toward maintaining the sample system and assuring that an adequate supply of reagents are on hand. Reagent consumption is typically 4 liters/month. Experience has shown that biweekly checks on these systems are warranted and that corrective action should be taken at that time if necessary. The biweekly check might include checking sample system flows, pressures and temperatures, and reagent levels; ensuring that the rotating reaction cell is operating correctly; and checking that no error messages are displayed on the digital panel meter. Error messages are seldom transmitted to any receivers, panel board instruments, or other data terminals. When the instrument diagnoses a problem, it will display an error message on its digital panel meter and hold its last good analog output so that process control upsets are minimized or do not occur.

8.1.1. Theory of Operation

Automatic titrators perform various chemical analyses. Functions are sequenced through a front panel keyboard or EPROM chip for sample fill, reagent dispense, spin, rinse, wait, and detector sensing actions to execute the desired titrametric, colorimetric, or specific ion analysis procedure. Precise measurements for different analytical applications are obtained through the use of the microprocessor and modular hardware configuration. Figure 8.1 shows four different sensors that may be used along with the rotary reaction cell and up to five digitally controlled burette assemblies. Dilution water may be added to the rotary reaction cell by means of a burette assembly or fixed-volume solenoid-operated valve (not shown).

Basic operation as an automatic titrator is as follows. The rotary reaction cell, shown in detail in Figure 8.2, is rotated slowly during most of the titration cycle. Sample, diluent, and various reagent burettes are sequentially filled through air-operated, sliding ceramic valve assemblies. After the burettes are filled, a small volume of liquid from each burette is dumped into the rotary reaction cell to assure that any air bubbles are expelled from the burettes. Water is then added to the cell and the cell is rinsed and emptied by using centrifugal force developed by spinning the cell at high speed. A second water wash and rinse sequence is used to remove any trace impurities remaining in the cell after the first wash and rinse. A small predetermined volume of sample is then added with the

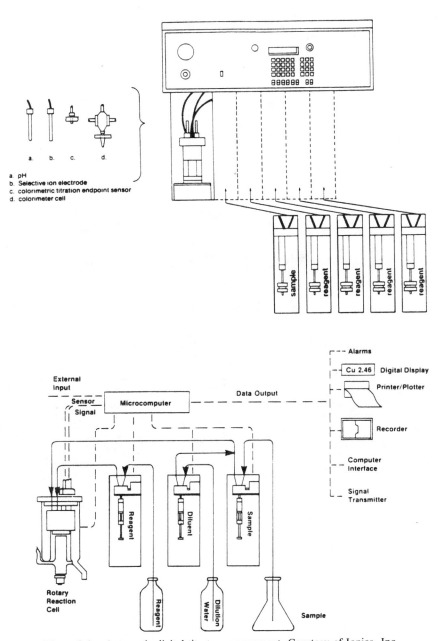

a. pH
b. Selective ion electrode
c. colorimetric titration endpoint sensor
d. colorimeter cell

Figure 8.1. Automatic digital titrator arrangement. Courtesy of Ionics, Inc.

189

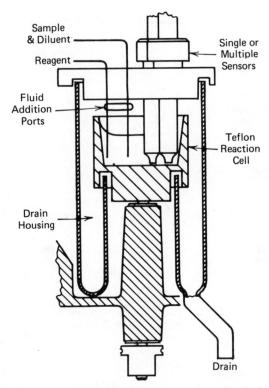

Figure 8.2. Rotary reaction cell of automatic titrator. Drive pulley is shown at the bottom of the illustration. Courtesy of Ionics, Inc.

reaction cell rotating slowly, followed by a predetermined volume of diluent. After waiting a few seconds to assure complete mixing, a titrating reagent is incrementally added in microliter quantities until a specified end point or pH level is reached. When the titration is complete, the output is displayed on a digital panel meter and held until the next titration cycle. The panel meter, printer, and/or recorder are updated at the end of each titration cycle. The wash, rinse, and reagent addition cycle is then repeated as programmed by the keyboard RAM memory or permanent PROM memory.

Figure 8.3 shows the printout for a typical titration sequence. The sequence shown can be programmed into memory or recalled from memory, edited, and changed. The microprocessor also provides prompting lines during programming and editing modes. In effect, the program shown in Figure 8.3 says:

SEQ #?
ØØ

Ø1 REPEAT SEQ 1Ø

COUNT Ø1

Ø2 REPEAT SEQ 12

COUNT Ø1

SEQ #?
1Ø

EDIT
Ø1 SEQ START 1Ø

Ø1 SEQ START 1Ø

Ø2 SPIN (Ø-99) 99

Ø3 RINSE Ø5

Ø4 SPIN (Ø-99) 5Ø

Ø5 RINSE Ø9

Ø6 RINSE Ø9

Ø7 FILL BURETTE Ø1

SOURCE? ØØ

Ø8 FILL BURETTE Ø2

SOURCE? ØØ

Ø9 SPIN (Ø-99) 99

1Ø WAIT (1-9999) SECS ØØØ5

11 SPIN (Ø-99) 5Ø

12 WAIT (1-9999) SECS ØØØ2

13 DISPENSE BURETTE Ø1

RATE? (1-99) 99

STEPS? (1-9999) 1ØØØ

14 TITRATE TO INFLECTION Ø2

TYPE? ØØ

RATE? (1-99) 9Ø

STEPS? (1-9999) 2ØØØ

SLOPE? Ø1

START PH 2.5ØØ

THRESHOLD Ø3

15 SEQ END 1Ø

SEQ #?
12

EDIT
Ø1 SEQ START 12

Ø1 SEQ START 12

Ø2 MATH EQ Ø1

K= 1.ØØØ

B= Ø.ØØØ

OUTPUT? Ø1

STORE AT Ø1

PORT Ø4

FULL SCALE = 2.ØØØ

ALARM PORT ØØ

Ø3 SEQ END 12

Figure 8.3. Typical computer printout of automatic titrator sequence. Courtesy of Ionics, Inc.

191

1. Start sequence 10.
2. Spin reaction cell at high speed.
3. Rinse with water for 5 s.
4. Spin cell at moderate speed.
5. Rinse with water for 9 s.
6. Rinse with water again for 9s for a total of 18 s.
7. Fill burette 1 from 1-liter reagent container.
8. Fill burette 2 from 1-liter reagent container.
9. Spin cell at high speed.
10. Wait 5 s.
11. Spin cell at moderate speed.
12. Wait 2 s.
13. Dispense 1000 steps of sample at a high rate.
14. Dispense titrant at a moderate rate until the pH signal first derivative peak is reached and then display voltage. ("Type," "rate," "steps," "slope," "starting pH," and "threshold" discrimination are titration instruction modifiers.)
15. End sequence 10.
16. Start sequence 12.
17. Solve math equation 1 where $X = KV + B$. (V is the number of steps of titrant; K, the scale factor; and B, the water blank constant. The terms "output," "store at," "port," and "full scale" are equation modifiers determining where the output will be transmitted and what the full-scale instrument range will be.)
18. End sequence 12.
19. Reset to step 01 and start over again when in automatic mode.

8.1.2. Calibration Techniques

Automatic titrators are generally calibrated against a known standard or actual process samples that have been analyzed by a good laboratory. A straight-line caustic calibration curve for the titrator should take the form of:

$$X = KV + B \qquad (8.1)$$

where K is the scale factor, V is the number of steps of titrant, and B is the water blank or number of steps of titrant required to neutralize the dilution water prior to any chemical reaction with the sample. B will be positive in caustic titrations when the water supply is slightly alkaline and negative in caustic titrations when the water supply is slightly acidic. Ion-exchange cartridges are used to provide the highest grade of pure water available to minimize the effects of water blanks. Other equations are used for other applications as listed in the applications section of this chapter.

To determine the optimum values of K and B, the initial values of $K = 1.0$ and $B = 0$ can be chosen and loaded into the microprocessor program so that actual plots of titrant steps versus caustic concentration can be obtained.

Two liters of each standard and sample shown in Table 8.1 were obtained, and the automatic titrator was programmed to titrate until a pH of 6 was obtained. All standards and samples were analyzed by an analytical laboratory to verify caustic, sodium hypochlorite, and salt concentrations. Twenty to eighty titrations were made on each standard and sample, and the raw count numbers or steps of titrant required to titrate to a pH level of 6 were reproducible to within ±0.35% of reading. Tests were run with and without the addition of hydrogen peroxide to neutralize the effects of sodium hypochlorite. Plots of these data are shown in Figure 8.4. After plotting this information, the actual values of B and K were determined as follows. Since the amount of hypochlorite in the process varies and substantially shifts the calibration curve, the instrument would have to be calibrated by adding hydrogen peroxide reagent to neutralize

Table 8.1. Automatic Titrator Calibration Data

Sample Number	Standard or Sample Concentration	Raw Counts or Steps of Titrant	
		With H_2O_2	Without H_2O_2
1	1.06% standard	123	124
2	5.04% standard	542	545
3	9.01% standard	1074	1080
4	1.2% NaOH 2.5% hypochlorite 4.2% salt	153	297
5	2.8% NaOH 2.5% hypochlorite 4.2% salt	350	491
6	4.6% NaOH 2.4% hypochlorite 4.3% salt	574	713

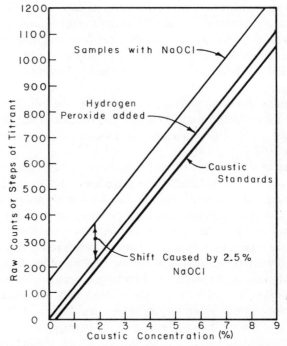

Figure 8.4.   Titrator calibration data for titrating a strong base with a strong acid.

the sodium hypochlorite. B is essentially zero for this curve. The K scale factor is then the actual concentration of the standard (or sample) divided by the raw counts obtained. For example,

$$K = \frac{2.8 \text{ (concentration)}}{350 \text{ (raw counts)}} = 0.008 \qquad (8.2)$$

The actual calibration curve for this particular process then becomes

$$X = 0.008V \qquad (8.3)$$

After a precise calibration has been obtained for a specific process material, the titrator will not read correctly on the standards unless the process curve and standard curve happen to coincide. In this case, the 1.06% standard read 0.95%, the 5.04% standard read 4.3%, and the 9.01% standard read 8.55% following instrument calibration on process samples. A plot of standard readings should be made for future calibration checks using the same or similar caustic standards.

8.1.3. Design Features and Applications

Figure 8.5 shows a block diagram of the automatic titrator. The heart of the unit is an Intel 8080 microprocessor mounted on the central processor unit (CPU) board. The rotary reaction cell assembly can accommodate up to three different sensors for multiple measurements on the same process sample. Each stepper burette board controls up to two burettes dispensing assemblies. Function boards such as the colorimeter board, air burette board, E/I output board, and RS-232 printer interface boards are optional. The optional D/A and E/I board is used for closed-loop applications where the titrator controls the final element such as a control valve. The RS-232 printer interface board is useful for troubleshooting the equipment and editing user-defined programs.

A summary of major design features is given in Table 8.2. It should be noted that instrument accuracy, repeatability, and response time vary widely and depend on the measurement being made. AC power, a 75-psi instrument air supply, and dilution water supply are required for operation. Required instrument airflow is about 50 cm^3/min, and about 8 gal of dilution water are used per month.

Sample temperature should be within the range of 1 to 40°C. Sample consumption is typically 10 cm^3 per analysis, and suspended solids of up to $\frac{1}{16}$ in. in diameter can be handled by the manufacturer's internal sampling system.

The self-diagnostics program is an important feature built into the automatic titrator. A brief listing of the error codes and their meanings follows:

Error 1 Keyboard entry mistake

Table 8.2. Summary of Major Design Features

	Titrimetric	Colorimetric	Selective Ion
Detection method	Derivative end point	Multiwavelength	Direct or standard addition
Range	≥1 ppm	≥0.01 ppm	≥0.05 ppm
Accuracy[a]	≥0.2%	≥3%	≥5%
Repeatability[a]	0.1%	1%	5%
Response time	≥3 min	≥6 min	≥1 min

Source: Courtesy of Ionics, Inc.
[a] A function of required response time, minimum maintenance, and sample conditioning.

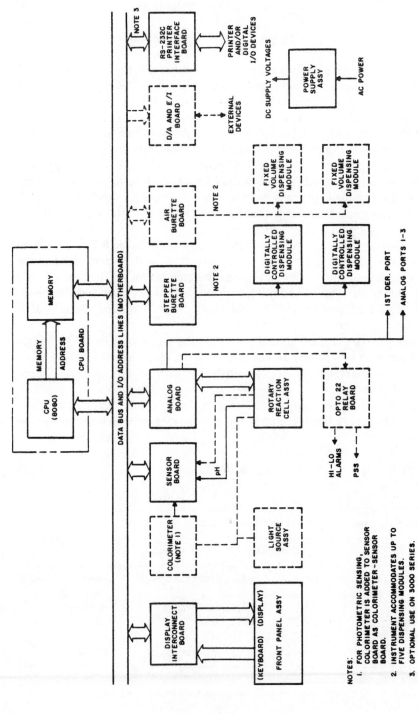

Figure 8.5. Functional block diagram of automatic digital titrator. Courtesy of Ionics, Inc.

196

Error 2	Printer is not accepting data and has caused titrator to stop
Error 3	Maximum titration count exceeded
Error 4-1	Burette 1 stroke limit exceeded
Error 4-2	Burette 2 stroke limit exceeded, etc.
Error 5	RAM failure
Error 6	Measurement result limit exceeded
Error 7-2	Sensor 2 input to A/D converter is less than 0.5 V
Error 7-3	Sensor 3 input to A/D converter is less than 10 V

As shown, the error codes can be used to troubleshoot the instrument. A full explanation of their meanings and the appropriate action to be taken is beyond the scope of this text. However, the built-in diagnostics enables the user to define a problem, correct it, and return the instrument to normal operation with minimum downtime.

Applications

A summary of design applications is given in Table 8.3. The table is divided into three main sections: titrimetric, colorimetric, and selective ion. A number of math equations have been programmed into the software of the automatic titrator and can be recalled from memory and used in the analysis program. The equation selected is chosen on the basis of the intended instrument application. Available math equations include:

For automatic titrations:

$$X = KV + B \qquad (8.4)$$

$$X = KV - B \qquad (8.5)$$

$$X = -KV + B \qquad (8.6)$$

$$X = -KV - B \qquad (8.7)$$

For colorimetric analyses:

$$X = KE \qquad *(8.8)$$

$$X = K \log SN \qquad (8.9)$$

Table 8.3. Design Applications

Titrimetric	Colorimetric	Selective-Ion
Types	Types	Type
Acid–base	Equilibrium	Direct concentration
Karl Fischer	Kinetic reaction	
Types	Automatic compensation	Features
Oxidation–reduction	for change in	Automatic standardizatic
Nonaqueous	Temperature	Temperature compensati
Back titrations	Source intensity	and control
Complexometric	Coating of optical	Automatic sample
Precipitation	surfaces	preparation
Amperometric	Optical transmission	Buffer or reagent
Methods	characteristics of the	additions to
Single end point	sample	Adjust pH
Multiple end point	Background sample	Adjust ionic
First derivative	color	strengths
detection	Sample turbidity	Mask interferences
Sensor-level detection	Sensor	Automatic incremental
Gran's plot	Colorimetric cell	methods
Sensors		Gran's plot
Potentiometric		Known addition
Colorimetric end point ᶠ		Known subtraction
		Sample addition
		Sample subtraction

Source: Courtesy of Ionics, Inc.

For specific ion measurements:

$$X = K10^{-SN} \tag{8.10}$$

where V is the number of steps or volume of titrant dispensed, E is the sensor output voltage, K is the scale factor, B is the water blank, and SN represents the equation solution storage location. Other software equations are available to the user for entering a constant into an equation solution, doing multiple end-point titrations and summations or differences, and providing automatic calibrations.

Figure 8.6 shows a typical sample system used for caustic titrations. The sample system features restricting orifices in the water and caustic lines to lower process pressures to an acceptable level for the plastic tubing used with replaceable filters, pressure gauges for monitoring, and ion-exchange columns for water conditioning.

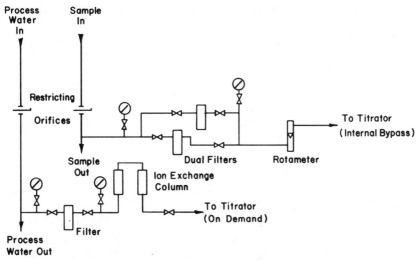

Figure 8.6. Typical sample system for automatic titrator.

8.1.4. General Comments

The automatic titrator described in this section is a precision metering device capable of making accurate and repeatable titrimetric determinations of the concentration of various acids and bases. It has been described as "a chemist in a box," but it is more than that; it is not subject to the human error and judgment of the laboratory analyst.

Improvements made over the past decade include:

1. Microprocessor control with user programming flexibility.
2. Incorporation of materials such as Teflon®, sapphire, and ceramics for improved service life.
3. Rearrangement of key electromechanical and mechanical parts such as burette drive assemblies and rotary reaction cell bearings to minimize their exposure to process materials.

Automatic titrators have progressed from the laboratory to process areas where they are being applied to an increasing number of tough problems involving on-line measurement and control. This trend is expected to continue as more and more laboratory instruments are adapted for and applied to the chemical and petrochemical industries.

8.2. DIGITAL DENSITOMETERS

There are a number of digital densitometers available today operating on the principle of a vibrating U tube, vane, or spool piece. These probe outputs are digital in nature because frequency is the fundamental variable measured. In its simplest form the electronics unit provides amplification, linearization, pressure correction, the necessary mathematical computations, and various output signal conversions.[2] When microprocessors are used in conjunction with flow measuring devices, the corrected density and mass flow of more than one component of interest may be computed. Four types of density-related outputs that may be provided are:

1. A frequency signal equal to the operating probe resonance frequency.
2. A 4- to 20-mA analog signal proportional to the density span.
3. A digital display of density directly in density units.
4. A BCD output signal that is linear with density.

8.2.1. Theory of Operation

The vibrating vane densitometer shown in Figure 8.7 consists of a vane positioned across the diameter of a supporting cylinder and operating at its resonance frequency to measure fluid density. The fluid may be either a liquid or gas. As the vane moves in simple harmonic motion it accelerates the surrounding fluid. In turn, the fluid inertially loads the vibrating vane, increasing its effective mass. As fluid density changes, the effective mass changes and the resonant frequency of the vane varies. The resonant frequency of the vane is inversely proportional to the density of the fluid in accordance with the following equation:

$$p = \frac{A}{f^2} - B \qquad (8.11)$$

where p is density, A and B are calibration constants, and f is frequency.

The resonant frequency of the vane is detected by a small piezoelectric ceramic sensor imbedded in the vane support ring. Vane vibrations are converted into proportional electrical pulses by the transducer and further amplified within the probe to improve the signal : noise ratio. The amplified probe signal is then sent to the transmitter, where it is amplified further and conditioned to provide a drive voltage for the magnetostric-

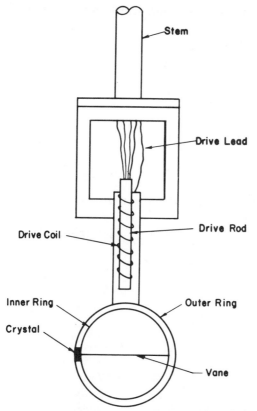

Figure 8.7. The vibrating vane densitometer. Courtesy of ITT Barton.

tive drive system that sustains the probe oscillations at the resonant frequency. Phase angle correction is built into the electronics unit to compensate for high-viscosity samples when needed.

Calibration constants A and B can be set so that the digital display reads directly in engineering units such as lb/ft^3, specific gravity units, or kg/m^3.

Another digital densitometer[3] uses a straight smooth-bore tube vibrating at natural frequencies from 1620 Hz to less than 1260 Hz to measure liquid density over the range 0 to 3000 kg/m^3. The frequency signal from the densitometer cell is fed directly into a microprocessor that computes process density without any loss in accuracy. Frequency inputs from density transducers are filtered and a set number of pulses are counted. The pulse duration time is measured and the time period of the input signal is calculated.

The heart of the vibrating tube densitometer is a microprocessor that accepts two periodic time density or specific gravity inputs, five analog inputs (three differential pressure, one line pressure, and one calorimeter signal), two pulse count turbine flow signals, and one temperature input. Because of the variety of inputs, gas and liquid densities can be measured simultaneously, or pressure–temperature-corrected mass flow measurements can be made. The keyboard–display arrangement in the microprocessor provides access for interrogation of the process and enables the user to take corrective action.

The versatility of the microprocessor is illustrated by the number of process equations it can solve.

For density calculation from time period:

$$D = K(K_0 + K_1 t + K_2 t^2) \qquad (8.12)$$

For density calculation from P, T, m, Z data:

$$D_d = \frac{K_m P}{(T + 273)Z} \quad \text{where } Z = F(TP_m) \qquad (8.13)$$

For temperature correction of density transducer:

$$D_T = DF(T) \qquad (8.14)$$

For pressure correction of density transducer:

$$D_p = D_T F(P) \qquad (8.15)$$

For referred density calculation:

$$D_R = D_c + F(DT) \qquad (8.16)$$

For API degree calculation:

$$API = \frac{141.5}{D_R} - 131.5 \qquad (8.17)$$

For flow rate from turbine pulse input:

$$V = Kf \qquad (8.18)$$

For flow rate from dP analog input:

$$V = K \left(\frac{dP}{D_c}\right)^{1/2} \tag{8.19}$$

For flow rate from a magnetic flowmeter:

$$V = KI \tag{8.20}$$

For temperature correction of flowmeter:

$$V_T = V[1 + K(T - 20)] \tag{8.21}$$

For pressure correction of flowmeter:

$$V_p = V_T[1 + K(P - 1)] \tag{8.22}$$

For mass flow rate:

$$M = V_p D_c \tag{8.23}$$

For standard volume flow rate:

$$V_s = \frac{M}{D_R} \tag{8.24}$$

For gas calorific value calculation from AGA 5:

$$C_m = K_1 - \frac{K_2}{m} \tag{8.25}$$

For energy flow rate:

$$E = MC_m \quad \text{or} \quad V_s C_v \tag{8.26}$$

Definitions

$F(\ )$	Function of
C_m	Calorific value (mass units)
C_v	Calorific value (volume units)
D	Density (uncorrected)

D_c	Density (temperature, pressure, and composition corrected)
D_d	Density (derived)
D_p	Density (pressure and temperature corrected)
D_R	Density (temperature referred)
D_T	Density (temperature corrected)
f	Turbine pulse frequency
I	Magnetic flowmeter signal
K	Constant
m	Molecular weight (D_R for gases)
M	Mass flow rate
P	Line pressure in bars
dP	Differential pressure signal
T	Line temperature in degrees Celsius
V_p	Volume flow rate (temperature and pressure corrected)
V_T	Volume flow rate (temperature corrected)

A simplified block diagram of the process computer is shown in Figure 8.8. Input and transducer power supply lines are not shown.

8.2.2. Calibration Techniques

Calibration of liquid densitometers is not particularly easy. Some liquids such as alcohol have large density changes over relatively small temperature ranges. In these applications precise temperature control or measurement and compensation is required. Air bubbles can become trapped on sensing cell surfaces, which will create lower than actual density readings. For these reasons the toxicity, density stability, and surface tension of the calibrating fluids must be considered. Water is an excellent calibration fluid except for its high surface tension. The air bubble problem can be minimized by using a small amount of wetting agent, which reduces the surface tension without significantly affecting liquid density. Entrained air can be removed from water by heating it to 180°F and allowing it to cool to room temperature.

When two or more liquids are used to calibrate a densitometer, the actual density of the liquids should be checked in the laboratory using a calibrated volumetric flask and an accurate analytical balance because handbook values of liquid density vary greatly depending on temperature and impurities. The densities of some common liquids are shown in Table 8.4. Solutions such as saturated salt solutions can also be used as calibrating liquids.

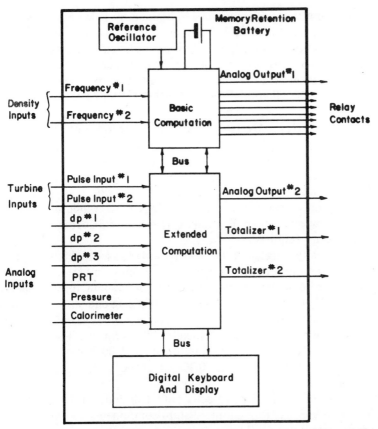

Figure 8.8. Functional diagram of density and mass flow computational system. Courtesy of Solartron Transducer Group, Jantech Corporation, 114 State Street, Boston, MA 02109.

Table 8.4. Partial List of Liquid Densities

Liquid	g/cm³ at 20°C
Acetone	0.792
Kerosene	0.820
Turpentine	0.859
Olive oil	0.918
Water	1.000
Sucrose	1.144
Glycerin (100%)	1.262

Rough calibrations are sometimes obtained by using water as a reference and observing the indicated process density changes and comparing them against laboratory samples. Various statistical routines can then be applied to the data to determine the approximate zero, span, and temperature compensation scale factors or settings. Twenty or more new data points can then be taken and the calibration settings fine tuned.

8.2.3. Design Features and Applications

Densitometers are designed to work with gas and liquid density cells and automatically compensate for pressure and temperature variations. Some control units accept flow sensor signals and provide mass flow or flow with density correction computations. Mass flow computations are becoming more popular in applications where material accounting or material balances are required.

Typical density ranges are 0 to 300 kg/m³ for gases and 0 to 1000 kg/m³ for liquids. Maximum density ranges of 0 to 400 kg/m³ for gases and 0 to 3000 kg/m³ for liquids are available from some manufacturers. Operating pressures and temperatures in excess of 2000 psig and 90°C are possible. With gases, temperature sensitivity is typically 0.002 kg/m³/°C; with liquids, temperature sensitivity is typically 0.02 kg/m³/°C. The density cell may be separated from the control unit by up to 2500 ft of cable. Accuracy with digital circuitry is about ±0.1% of range with a repeatability of ±0.02% of range.

Many gas and liquid density cells mount directly in the pipeline or a sidestream; therefore, few sample system components are required. Digital densitometers are used to:

1. Measure the density or concentration of various dissolved or suspended solids, liquids, or gases.
2. Measure the density of high-pressure natural gas and other hydrocarbons such as ethylene.
3. Measure or determine gas quality analysis for gas metering stations.
4. Measure the percentage of oil in water over the range of 0 to 5%.
5. Identify products or detect the interface in a multiproduct pipeline system to prevent product contamination.[4]
6. Calculate the mass flow rates of dissolved or suspended solids, liquids, or gases.
7. Determine jet or diesel engine performance by measuring the pounds of fuel consumed to obtain various thrusts.[5]

8.2.4. General Comments

Density measurement and control is often considered another form of process measurement and control as contrasted to an on-line analyzer measurement. However, when density measurement is regarded as a method for determining the composition or concentration of a chemical compound, its consideration as an analytical measurement is valid.

Modern materials such as tantalum, titanium, and tough ceramics are now used so that process wetted parts are virtually immune to the corrosion of most chemical processes. Also, the use of direct insertion probes has minimized pluggage in slurries and high-density, high-viscosity streams. New probe materials and designs have elevated the densitometer above the "for clean streams only" category. Factors affecting density such as entrained air bubbles or fine solid particles are still of concern when these are not the variables of interest. In many of these cases changes in the piping configuration and good sample conditioning can eliminate or help to minimize these problems.

All the various forms of vibrating densitometers are expected to eventually take advantage of microprocessor features to improve their usefulness, flexibility, and accuracy. The ability to make mass flow measurements is a feature that is expected to receive continued interest from both process control and instrument engineers.

8.3. OCTANE ANALYZERS

A good example of a "dedicated" on-line analyzer system using a microprocessor is the octane analyzer.[6] This type of analyzer is designed for continuous unattended operation and ease of maintenance. User access is of secondary importance and consists of inputing scale factors for precise calibration of engine-rated octane standards and initiating test routines for diagnosing analyzer operation. The octane analyzer is designed to be field mounted near its sample take-off point. Key variables are sent from there to a remote printer. The microprocessor controls the gasoline injection sequence, accepts temperature signals, and computes the research octane number (RON). The remote printer prints out operating temperature, RON number, and a RON bar chart. The octane analyzer can provide advisory information for manual control of gasoline refineries and pilot plants, or the unit can be placed in closed-loop control for automatic process control. The octane analyzer does only one job, but it does it well; it measures and controls the RON at various manufacturing phases in the gasoline refinery.

208 DIGITAL ANALYZERS

8.3.1. Theory of Operation

The octane analyzer simulates "cool flame" partial oxidation reactions that occur during engine measurements to determine the RON of gasolines. The cool flame reaction measured occurs prior to combustion. The process gasoline sample flows to an air-actuated pump that pressurizes it to 60 psi to prevent vaporization in the sample lines. At 5-min intervals a 12-μl gasoline sample is automatically injected into air flowing into a heated reaction chamber. Air for the reaction is preheated to $\pm1°C$ in a temperature-controlled oven and flow controlled at 80 ml/min. The air serves as a carrier gas for the liquid gasoline sample prior to its entering the reaction chamber, which is temperature controlled to $\pm0.1°C$.

Partial oxidation takes place after an initial delay, and the temperature rises and then falls, producing a peak similar to the one shown in Figure 8.9. The reaction is self-initiating and self-extinguishing. Both the intensity of the peak or peak height and the delay time or induction period provide a good correlation with octane number. As octane numbers decrease, reactions become more severe and the induction periods shorten.

The microprocessor correlates the RON number from the peak height and induction period and provides a printout of time, temperature, and RON along with a bar graph presentation of RON as shown in Figure 8.10.

8.3.2. Calibration Techniques

Automatic standardization can take place every 24 h or on demand. Calibrations can be made at any time using engine-rated samples. Usually the octane ratings of three known standards are entered into the microprocessor so that the printed output value closely tracks the engine readings. The octane analyzer operates unattended under normal operating conditions.

8.3.3. Design Features and Applications

The range of the octane analyzer is field adjustable, with a typical 5 octane number linear span. Repeatability is 0.1 octane number with the accuracy of calibration dependent on the accuracy of the engine-rated calibration standards. Dual-range analyzers are available with linearity specified as 1% at 1 RON from midrange and 5% at 2.5 RON from the midrange value. Response time is 60 s after injection of the sample. The water-free gasoline sample consumption is about 300 ml/h with maximum sample pressure to 300 psi at operating temperature limits of 10 and 40°C.

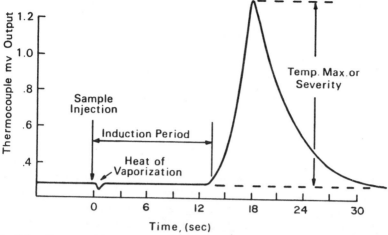

Figure 8.9. Curve used for determining research octane number (RON). Courtesy of Foxboro Analytical.

One option permits a printout of "fuel alarm" or "air alarm" if either of these services are interrupted or reduced. An optional "bad data" switch can be provided to permit manual interrogation of the microprocessor to alert the user of loss of analyzer power, loss of reactor air pressure, loss of sample pressure, excessive temperature deviation, or "off-line" sample analyses. Current outputs of 4 to 20 mA or 10 to 50 mA

TEMPERATURE RANGE 1 OR 2?
OPTION> 1

	02/23/78 BLOCK TEMP	OCTANE RON	95.0 +	96.0 +	97.0 +	98.0 +	99.0 +	100.0 +
47	316.7	99.9	***
52	316.6	99.9	***
57	316.6	99.9	***
2	316.6	99.9	***
7	316.6	99.9	***
2	316.6	94.9	<
7	316.6	95.2	***
22	316.6	95.1	**
7	316.6	95.1	**
2	316.6	97.7	**************************
7	316.6	97.8	**************************
2	316.7	97.8	**************************

ure 8.10. RON graph presentation shows analysis time, block temperature, and RON trend. urtesy of Foxboro Analytical.

are available, or voltage outputs of 0 to 1, 0 to 4, 0 to 5, 0 to 10, or 1 to 5 V DC are available depending on user selection and requirements.

The octane analyzer is designed for use in Class 1, Group D, Division 1 atmospheres but must be sheltered from the weather. A glass reactor is used for leaded fuels, but metal reactors are used for nonleaded fuels.

8.3.4. General Comments

Sections 8.1 and 8.2 of this chapter dealt with digital analyzers that were extremely flexible in that user access could be achieved through built-in keyboards or programming switches. In the case of the digital titrator, up to 100 user-defined programs can be entered into RAM memory. In the case of the octane analyzer, only calibration data can be placed into program memory. There are advantages and disadvantages to both design concepts. When user access is a prime feature, the user must be familiar with the terminology of the machine; this may not be easy or natural for everyone. Once the terminology is understood, anyone with access to a programming key can gain access to the software and modify or rewrite it so that some form of key control is usually instituted. There are many applications where dedicated control is adequate and desirable, and in these cases complex internal computations must usually be preprogrammed into the analyzer by the instrument manufacturer.

CHAPTER

9

ON-LINE PROCESS CHROMATOGRAPHS

W. C. WELZ, JR.

One of the best analytical methods available for the determination of various components in process samples is molecular separation by chromatography. The first on-line chromatographs were crude compared to today's sophisticated, computer-controlled analyzers. As in the case of many other analytical laboratory instruments adapted for continuous process duty, the application of chromatographs to field installations was a result of the need for quicker and more specific stream analysis. Early processes were seldom much more than batch units or relatively simple distillation units that could be easily controlled by monitoring only a few process stream parameters. Modern process designs are becoming much more complex. Continuous operation of large process systems at maximum operating efficiency has resulted in the need for responsive process control information.

Time-consuming analytical methods are no longer adequate because of the need for more frequent data input to computers or fast-acting control systems. Digital multiplexing techniques and data manipulation have revolutionized modern process control. Precise analytical measurements of process stream chemical compositions is typical of the data that now must be input to computers for determination of process operating setpoints. Chromatographs are being used for composition measurements in other applications that involve chemicals; food processing, pharmaceutical manufacturing, and environmental pollutant detection are only a few examples. Chromatography also has applications in outer space for the determination of interplanetary soil and atmospheric composition. An example of this is the organics analyses performed aboard the Viking Landers on the planet Mars.

This chapter is oriented toward the typical on-line process gas chromatograph (GC). The same fundamentals described for process GCs can be applied to most other types of chromatograph.

211

9.1. CHROMATOGRAPHIC SEPARATIONS

Chromatography is a technique used to separate the volatile components of a mixture by distributing each one differently between two phases. One phase is a stationary liquid or solid, and the other is a mobile carrier that flows through the stationary phase. The distribution of a component between phases depends on interactions that occur between individual component molecules and stationary phase molecules based on molecular size, chemical activity, and component vapor pressure. The stationary phase selectively retards component molecules and collects or groups them into distinct narrow zones as they are transported through it by the mobile phase.

This technique was introduced around the start of the twentieth century.[1,2] Glass columns packed with activated charcoal, diatomaceous earth, or molecular sieves were also used with gravity-fed mobile phases in some of the early chromatographic separations. Modern gas–liquid chromatographic development did not occur until the 1950's.[3,4] Gas–liquid chromatography, usually abbreviated simply as GC, refers to the use of a column packed with very small, chemically inert, solid particles of particular size that have been coated with a very thin film of solvent. The solvent is chosen for its specific affinity for components of interest in the sample. The liquid-phase solvent selectively retards, separates, and even sometimes retains various components in the sample as carrier gas moves the sample through the GC column. Column diameters, lengths, operating parameters, and packing materials depend on the chemical characteristics of the sample and the required component separations.

9.2. BASIC GC SYSTEMS

Figure 9.1 is a simplified diagram of the main elements of a basic, on-line process gas chromatograph. Each part is discussed in detail, and any variations to it will be shown as the text continues. These elements comprise subsystems that are:

1. Carrier gas and utility supply system.
2. Sample injection system and programmer.
3. Chromatographic column(s).
4. Temperature-controlled oven enclosure.
5. Component detector.
6. Detector signal conditioning network.

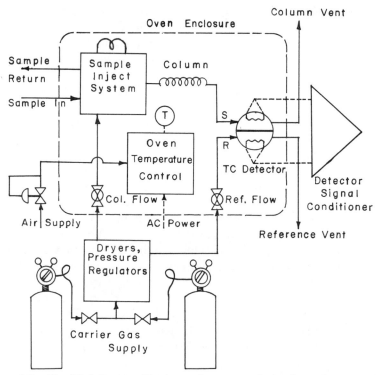

Figure 9.1. Simplified diagram of basic gas chromatograph showing major components.

9.2.1. Carrier Gas and Utility Supply System

The carrier gas and utility supply system provides all the services required for continuous analyzer operation. Most GC's require dry instrument-grade air, uninterruptable AC power, environmental protection, and low-pressure vent connections. Dry instrument air is piped to each GC to provide the motive force for the internal sample injection, backflushing, or column switching valves in the analyzer. This air is also piped to the temperature-controlled heater assembly in the oven to assure uniform heating of all devices inside the enclosure. Electrical subsections of the analyzer may require continuous air purges to provide protection from explosive gases or corrosive materials if the GC is installed in a hazardous environment. Many GC applications require analysis of materials that can become dangerous in the event of a power supply failure. Chemicals that might decompose or in any way damage the internals of valves, columns, or the detector require special consideration for their handling during utility supply outages. Deenergized valve positions should be specified to make the GC revert to a carrier gas only mode to keep samples or backflushed fluids out of the analyzer during utility failures.

Most chromatographs require environmental protection to enhance the temperature stability of their electronic circuitry. Small installations with only a few chromatographs located at widely differing areas cannot usually justify the fabrication of buildings or enclosures to house them. Such situations require extra attention to ensure that the analyzers have adequate protection from rain, direct sunlight, wind, chemical fumes or splashes, and lightning. A preferred installation is a centrally located shelter with environmental control. Additional service features such as positive air pressure to exclude hazardous fumes, communications networks, work areas, and emergency spare parts storage should be incorporated into the design of these shelters. Many sample systems incorporate heat tracing, purge and vent headers, high-flow or fast-bypass filter systems, and overpressure relief and flow-limiting devices that can also be centrally located on the analyzer shelter outside walls. Additional pressure regulators, flow-limiting devices, loss-of-carrier gas pressure switches, and on-line molecular sieve driers are often incorporated into the carrier gas tubing system inside the analyzer shelter. This ensures that a constant supply of clean, dry carrier gas is available to each GC. The tubing used in these systems should be acid washed, solvent flushed, and dried with hot inert gas to remove contaminants that might adversely affect GC performance. Tubing manufacturers follow several different procedures to produce chromatographic grade tubing, and special care should be taken to insure that their cleaning steps are adequate and compatible with the intended tubing application.

Carrier gas cylinders and calibration gas cylinders are usually placed along the outside shelter wall in cylinder storage racks. These racks should have retaining chains and room for temporary storage of a few full or empty cylinders. When empty gas cylinders are replaced with full cylinders, the pressure-reducing regulators used for that particular gas service should be reinstalled on the same service to avoid the possibility of cross-contamination. Two supply cylinders with individual pressure regulators are usually required for continuous GC operation. The delivery pressures are slightly offset so that one cylinder stays on line while the second remains at full pressure until the first is expended. All personnel that come in contact with high-pressure, compressed-gas cylinders should be thoroughly trained to use the storage and handling procedures recommended by specialty gas manufacturers. The Compressed Gas Association[5] has developed standards that specify different and noninterchangeable regulator to cylinder valve connection devices called CGA fittings. These fittings prevent regulator connections to cylinders that would mismatch fuel and oxygen, toxic and nontoxic, or corrosive and noncorrosive services.

Full carrier gas cylinders usually have over 2000 psig, and full calibration gas cylinders may also contain this pressure or as little as 500 psig, depending on the vapor pressure of the cylinder's contents. Analyzer maintenance technicians should check carrier gas cylinder pressure daily and consider changing those cylinders with pressures of less than 100 psig. It is not good practice to allow a carrier gas cylinder to completely expend itself because of the possibility of water vapor or hydrocarbon contamination caused by the migration of these vapors out of the interior cylinder walls. Some users may prefer to expend the contents of the cylinder down to a lower pressure if the application can tolerate possible carrier contamination during times of reduced cylinder pressure. Most of the regulators used for carrier gas have cylinder pressure gauges with a range of 0 to 2500 psig. It should be kept in mind that a 100-psig indication could sometimes be a 4% error in the gauge reading. The possibility of analyzer downtime or damage is not worth the cost of the unused carrier gas in cylinders taken off line at 100 psig. There is also a possibility of air ingestion through a cylinder valve that was left partially open after the cylinder was removed from service. This is a special hazard with hydrogen, methane, or any other flammable carrier gas. Many calibration gases can be toxic, corrosive, or susceptible to degradation unless special cylinder materials or handling procedures are used. Carrier gas quality is sometimes critical because of the effect contaminants such as air, oil, or water vapor can have on chromatograph performance. Chromatograph-grade carrier gas cylinders have special identification marks so that these cylinders can be refilled with exactly the same gas every time.

Specialty gas manufacturers should be required to provide information pertaining to contaminants that might affect chromatograph calibration or operation in addition to recommendations for maximum cylinder shelf life. Analysis of trace quantities of hydrogen or light hydrocarbons sometimes requires ultra-high-purity carriers because the amount of hydrocarbon contaminants in lower grades may exceed the range of the required analysis. Corrosive materials are especially subject to changes in composition over relatively short times compared to inert materials because of chemical interaction with the internal components of the cylinder. Calibration gas cylinders containing hydrocarbons with high boiling points may sometimes require thermal jackets or heating coils to vaporize these components during cold weather.

9.2.2.   Sample Injection System and Programmer

The volume of sample injected into the column is usually very small compared to the volume of the tubing used to bring the sample to the injection valve. Fast-flowing, high-bypass filter loops can be used for reduction of the sample residence time in long sample tubing runs. Two varieties of these filters are shown in Figure 9.2. The bypass side of the filter system will return to the process while the filtered effluent is piped to the analyzer. If the sample is at first liquid, it is vaporized prior to entering the chromatograph. Care must be taken to avoid flashing and unwanted component condensation during pressure reduction if the incoming sample is at relatively high pressure.

Chromatographs used for laboratory applications rarely have fixed-volume injection valves because these analyzers have been designed to be adaptable for various separations. These laboratory GC's have septums that are pierced during sample injection with the needle of a gastight syringe filled with the correct sample volume. However, on-line process GC's are specified for exacting sample conditions and repetitive component separations. Most process GC's use $\frac{1}{8}$-in.-OD (outer diameter) or $\frac{1}{4}$-in.-OD columns less than 10 ft long and sometimes less than 1 ft in length. Sample volumes can range from less than 1 μl to several milliliters. The volume of sample introduced into the column during sample

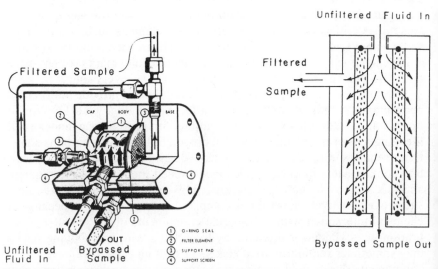

Figure 9.2.   Two types of high-speed bypass filter. Left illustration courtesy Collins Products Company.

injection depends primarily on the physical characteristics of the sample, the size of the column, and the amount of liquid-phase solvent used on the column packing.

It is extremely important that the sample injection valve introduce exactly the same quantity of sample during the start of each analysis cycle regardless of the sample volume required. Precision control of column operating parameters such as temperature, pressure, and carrier gas flow rate are just as important for reasons that are discussed later during the examination of chromatographic column theory. Almost all injection valves use fixed-volume cavities (internal or external) for precise reproducibility of sample volumes. Sample block and vent valves are often used inside the GC oven just ahead of the sample inject valve to stabilize sample pressure in the injection valve immediately before the sample is introduced into the column.

One of the most popular methods of sample volume repeatability utilizes a selected length of small-diameter tubing connected to the injection valve as shown in the simplified flow diagram in Figure 9.3. The sample volume loop can be seen connected to valve ports 4 and 9. Dashed lines shown inside the rectangle indicate the porting arrangement of the inter-

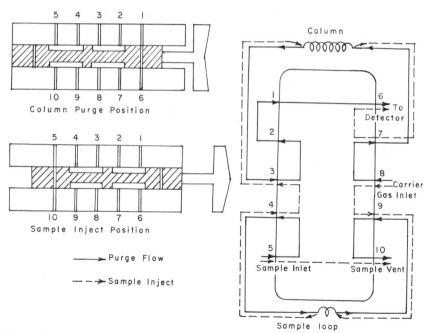

Figure 9.3. Illustration showing sample injection valve and flow paths during sample inject and purge periods. Courtesy of AMSCOR.

nal valve mechanism during the short time that the mechanism is shifted into the sample inject mode. Solid lines indicate the flow paths through the valve at any time other than sample inject. Exact lengths of tubing can introduce precise amounts of sample because the volume of the tubing remains constant. A 1-ft length of 0.020-in. wall diameter, $\frac{1}{8}$-in. stainless steel tubing will typically have an internal volume of about 1 ml. Much smaller volumes can be obtained by using a very small, wire-gauge-size drilled cavity inside the internal mechanism of the injection valve.

Similar flow switching can be achieved with other configurations of sliding plate or spool valves. A somewhat different approach incorporates the use of plungers to alternately depress or release a thin-sealing diaphragm at various points around equidistantly spaced ports in a circular valve body. A simplified diagram of these flow paths is shown in Figure 9.4. By operating the plungers in opposing groups, the thin-sealing dia-

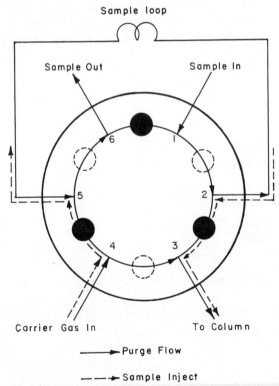

Figure 9.4. Six-port plunger-type sample injection valve with flow paths shown during sample inject and purge periods. Courtesy of Applied Automation, Inc.

phragm can be deformed to open or close flow passages between adjacent ports located in the surface opposite the plungers. The manufacturer advertises a million cycle lifetime on clean fluids with operation at pressures of up to 1500 psig. The dashed circles indicate plungers withdrawn from the sealing diaphragm, thereby allowing flow between ports 1 and 2, through the external sample loop into port 5, and then out port 6. Carrier gas is routed during this time into port 4, past another retracted plunger to port 3, and from there out to the column. When the valve is energized or shifted into the sample inject mode, the two sets of three plungers quickly change their released or depressed positions. This changes the deformation of the sealing diaphragm, causing the previous flow paths to be closed and opening paths that were shut earlier. Reexamination of Figure 9.4 with this in mind reveals that the carrier gas flow can be seen to follow a different path as it passes from port 4 to 5, sweeps the sample injection volume into port 2, and then out port 3 into the column.

Figure 9.5 shows another circular valve with the ports switched by a

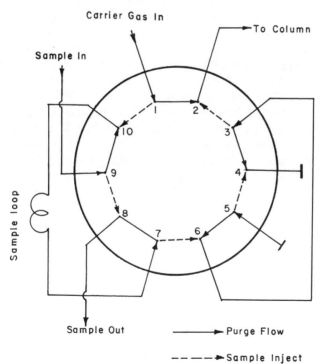

Figure 9.5. Ten-port rotary valve with flow paths shown during sample reject and purge periods. Courtesy of Analytical Instruments Corporation.

rotary sliding device instead of a reciprocating mechanism. A similar flow legend is used here where solid lines indicate flow during the deenergized time of the valve and dashed lines show flow when the valve is shifted into the sample inject mode. The flow paths between ports would be changed during rotation of the valve mechanism to the sample inject position, thus allowing carrier gas to enter port 1, flow over to port 10, and push the sample volume into port 7 and then over to port 6. From there the external tube connected between ports 6 and 3 would carry the sample volume into port 3, where the dashed line shows the rotating mechanism had connected that port to port 2 and from there out to the column. After this relatively brief sample inject period, all those previous flow paths would revert to the route shown by the solid lines as the rotating mechanism shifted back to its deenergized position. Sample would continuously flow through the external sample loop, and carrier gas would purge and regenerate the column as the analyzer either completed separation of the sample of the last cycle or waited for another analysis cycle to commence.

Interport leakage caused by very minute scratches on the sliding mechanism is a major problem with any valve that has to reproduce exact sample volumes many thousands of cycles. The scratches occur when very fine but abrasive particles are carried into the valve ports from either the sample system or the carrier gas system. It is for this protection from valve damage that high-quality analytical filters capable of removing particles as small as 1 μm are highly recommended by analyzer system designers. Filter mediums with this small pore size are almost always bypass types rather than full flow types because of the need to remove larger-diameter particles that would soon plug the full-flow filter medium. Additional roughing filters are often installed upstream of these small pore filters to enhance the service life of the high-bypass units. Porous sintered-metal filters, available from many tube fitting suppliers, can be installed in the sample line prior to routing this line to the high-bypass filter of the analyzer. Even with adequate filtration, sliding plate valves can develop interport leaks from chemical attack of their metal sealing surfaces or gradual wear of the slider block. Teflon® is the most commonly used sealing material for slider faces. Sometimes the complete valve assembly will be coated with it if the process stream is extremely corrosive or highly reactive. The sealing surfaces between ports must be maintained extremely flat and polished to avoid any small flow paths, channels, or voids. Lapping blocks are used to make all contact surfaces optically flat during valve overhaul. The plunger sealed diaphragm valve does not have to contend with these problems and is becoming increasingly used by chromatographers in many GC applications.

All sample injection valves should introduce the required sample volume into the column quickly as a slug or very narrow zone of material instead of allowing that volume to be mixed and diluted with the carrier gas. This is because the time it takes any component to elute from a column cannot possibly be less than the time it takes the complete sample volume to enter the column. If a sample of material was slowly introduced into the column by dilution with carrier gas, the column could not retard a very narrow band of molecules. The molecules of interest would be entering the front of the column over a long period of time and, even though the column's separation mechanisms would selectively retard those molecules, would require as much time to exit the column as the time required for their complete entry. To better understand the molecular grouping that occurs in chromatographic separation, Figure 9.6 should be referred to. This diagram represents a cross section of sample being separated as it is carried through a column. Each successive time interval shows the sample being separated into two main groups of molecules. The group that reaches the column exit first consists of molecules that tend to remain in the mobile phase. The second group is delayed because it has some

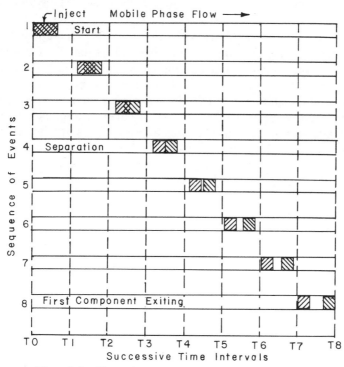

Figure 9.6. Illustration of two-component GC separation.

tendency to interact with molecules of the stationary phase. These interactions require equilibrium time, and this slight delay forms separate bands or zones of molecules that are retained according to the strength of the interacting forces.

Figure 9.1 shows only one column in the basic thermal conductivity GC, but many other configurations are often used in different applications. Process gas chromatographs may require multiple types of detectors, several more columns and valves, or even two different carrier gas streams to rapidly analyze complex samples. A particular analysis might require the column and valving configuration to rearrange the order of component elution sequence or prevent certain types of components from damaging the detector. Analysis cycle time can often be reduced in applications requiring the separation of only a few components from a complex sample by venting or backflushing those not of interest after obtaining those required for the analysis. Examples of this are the analysis of only the light components through C_4 hydrocarbons in gasoline streams or stripping acids or other undesirable components from other sample components that must be detected.

"Stripping," "precut," "heart-cut," "foreflush," and "backflush" are descriptive modifiers of words pertaining to column function or column and valve configurations for different GC separations. The required analysis may often require multiple valves that must be precisely switched at very definite times during each analysis cycle. The analysis cycle for most on-line process gas chromatographs begins with the event of sample block and vent or sample inject. Any later valve switches or changes in GC operation (detector signal balance, attenuation, integration, or data presentation) are initiated at preset times by a device called a *programmer.* Synchronous motor-driven cam and microswitch assemblies were used before logic circuitry was developed and incorporated into accurate timing devices like crystal-controlled oscillators. Microprocessors are becoming increasingly used in chromatographs not only for program control, but also for data conversion and communications interface control. Many of these digitally controlled chromatographs have extended capabilities to control more than one analyzer or communicate directly with a host computer system.

9.2.3. Chromatographic Columns

The chromatographic column is often considered the essence of the applied science used in the design of modern gas chromatographs. Numerous works have been devoted to sophisticated mathematical models that describe the specific operating parameters involved in columns.[6-9] Readers are encouraged to pursue these and other excellent sources available from industry journals, professional societies, and chromatographic equipment suppliers. This discussion is limited to a nonmathematical treatment of column function and attempts to give a more generalized view of column features.

Carrier gas is used to transport the vapor-phase components through the column. Process gas chromatograph columns of less than 10 ft in length and $\frac{1}{8}$ or $\frac{1}{4}$ in. in diameter are usually operated with carrier gas flow rates of 20 to 80 cm^3/min. The carrier gas pressure necessary at the column inlet to achieve this flow is a function of the column's resistance to flow and its outlet pressure at the detector vent. Column vents usually open to atmosphere to fix one end of the column at a constant absolute pressure so that good pressure regulators can maintain consistent flow rates. Some of the factors that influence column permeability are:

1. The internal diameter and length of the column tubing.
2. The size of the particles used as the stationary phase or as the support medium for the liquid phase.
3. The amount of liquid solvent coated on the solid support particles.
4. The molecular weight of the samples.

The last two factors primarily affect the time required for equilibrium to occur between component molecules traveling in the mobile phase and those interacting with the stationary phase. Even though interaction equilibrium has little effect on the physical characteristics of a column, it certainly influences the desired component separations. Several other physical and chemical properties of the column and its packing are involved in the separation of component molecules, and these also influence the separation equilibrium time. The viscosity of the liquid phase and its temperature, the strength of chemical bonding activities between solute molecules and molecules of the column walls, solid support particles, and liquid phase are all influential separation factors.

The primary methods used to obtain chromatographic separation of component molecules are molecular size exclusion, chemical adsorption, and solvent partitioning. Molecular sieves are capable of separating some compounds by molecular size because the crystalline structure of the sieve acts like a maze of closely sized pores that allow passage of small molecules and block passage of large ones. Consider a hypothetical sponge that absorbs only molecules of certain sizes; those that fit into the sponge pores are retained while the nonfitting molecules are not. Some separations are achieved by adsorption or the selective bonding of component molecules to the stationary phase by chemical attraction. Diatomaceous earth, activated charcoal, silica gel, and the newer amorphous polymers have active surface areas with specific affinities for certain types of molecule. Loose chemical bonds are formed between these active sites and some of the molecules of the sample. This action results in retention of groups of molecules based on the strength of adsorption. Surface activity on solid support materials intended for use in solvent partitioning columns can often be troublesome if component retention is influenced to the point that solute molecules are retained after partitioning equilibrium has been established.

Partitioning columns use solid supports that have been treated to neutralize active surface sites so that only the solvent coating is involved in the retention and separation of components. The solvent should have low vapor pressure, be similar in chemical structure to the components being separated, and be thermally stable. The selectivity of a solvent for a particular solute depends on the interaction forces that occur between the two molecules. A general statement that "like dissolves like" can be made about solvents in organic chemistry. Molecules that have electronegative elements such as fluorine, chlorine, oxygen, or nitrogen combined with carbon are usually polar and can be dissolved in polar solvents. Straight-chain hydrocarbons dissolve into nonpolar solvents. The distribution of solute molecules between the carrier gas and the stationary solvent is also a function of solute vapor pressure. High-vapor-pressure (low-boiling-point) solutes tend to remain in the carrier or mobile phase and will be eluted before low-vapor-pressure solutes that remain in the solvent longer. The establishment of equilibrium between vapor and liquid phases is also dependent on the interaction forces caused by hydrogen bonding or molecular polarity. A major advantage of gas chromatography is the ability to separate compounds of similar boiling point by using the molecular affinity of solvent for solute components. The time required to establish solute partitioning equilibrium can be influenced by the temperature of the column and the velocity of the carrier gas or mobile phase flowing through the column.

The amount of solute per unit volume of liquid phase divided by the amount of solute per unit volume of mobile phase is called the *partition coefficient*. Effective separation is usually possible if the components have different solubility ratios or partition coefficients. The partition coefficient is temperature dependent in that higher temperatures tend to cause solute molecules to spend more time in the vapor phase and less time in the liquid phase. Therefore, less time is required for those molecules remaining in the mobile phase to emerge from the column than those molecules remaining soluble in the stationary phase. Many variables influence column performance. Oven temperature, carrier gas pressure and flow, carrier gas purity, and sample injection volume are parameters that must be rigorously controlled for reproducibility of component retention volumes.

9.2.4. Temperature-Controlled Oven Enclosure

The sample injection valve, any column selecting valves, the columns and their flow control valves, and the component detector are all located inside the GC oven. This enclosure is usually aluminum or stainless steel lined, heavily insulated, thermostatically controlled, semiairtight box with a front-hinged door for access. Any GC electrical devices should be mounted in other purged enclosures. Low-mass air heaters are used inside the oven to recirculate hot air at a very specific temperature controlled to within ±1°C. This stabilizes the temperature of all internal oven components. Section 9.2.3 described how column temperature was a very critical parameter influencing the effective separation of sample components and thereby changing their subsequent elution times. Another primary concern about oven temperature control is that most liquid phases must be operated within a very definite temperature range. Many of these liquid phases have high viscosity at low temperatures or do not liquefy until the column has obtained some minimum operating temperature. The stationary phases can be chemically changed or permanently damaged if their upper temperature limit is exceeded. The solvent used to dissolve the chemical compounds that formulate the liquid phase can also be forced to slowly evaporate or bleed out of the column. Column bleeding can cause changes in column and detector performance or even result in the detection of false component peaks.

Not only is the liquid-phase temperature of concern to the chromatographer, but often the fluids being sampled must be maintained within definite temperature limits. High-boiling-point hydrocarbons may condense in columns or valves developing tars and even totally plugging valve ports if oven temperatures in these analytical systems are allowed to become too low. Polymerization of sample components can sometimes be a problem with certain chemicals if their temperature is allowed to become too high.

Almost all GC ovens have AC power interlocks to protect the low mass air heater elements against burnout if the heater air source fails. A temperature sensor is often included in the programmer control circuitry so that the analysis cycle is inhibited unless the oven is within prescribed temperature limits. Thermal conductivity (TC) detectors are very temperature sensitive and have a great influence on chromatographic stability.

Most TC detectors are housed inside a large metal block or cell so that temperature variations at the internal elements of the detector can be dampened by the large heat sink provided by the mass of the detector block.

Column temperature exercises a very definite effect on GC separation, and this effect can be useful in the separation of components having widely differing boiling points. Isothermal GC is useful for the separation of sample components within a boiling range of 100°C. Components having wider boiling ranges of up to 400°C can often be separated by programmed-temperature gas chromatography (PTGC). PTGC is used more in laboratory applications than in process gas chromatographs, however, because of the limited number of applications requiring continuous, on-line analysis of wide-range boiling-point samples and the somewhat more complicated chromatograph configuration necessary for the separate control of injector temperature, column temperature, and detector temperature.

Access to oven-mounted components can be readily obtained by opening the door, but this usually upsets the thermal stability of the analyzer, sometimes for hours. The small needle valves that are used to adjust the carrier gas flow rate through either the column or the detector must also be located inside the oven to minimize the effects of ambient temperature changes on these flows. GC manufacturers usually provide screwdriver-access ports in the door face that are aligned with slotted heads on the needle valves so that slight adjustments can be made to the valves without opening the oven door. Bubble burettes connected to the detector vents are usually used to measure these flows. External gauges and dial thermometers should be marked to indicate the normal operating condition of a chromatograph. This technique allows faster determinations to be made of chromatograph operation during maintenance routines or troubleshooting. On-line process GC oven temperature, carrier gas pressure, and air supply pressure should be monitored daily.

9.2.5. Component Detectors

Chromatographic detectors develop the signals necessary for the display of analogs that represent the elutants of the column. These detectors are differentiating types as opposed to integrating detectors such as titration burettes. Components eluting from the column are differentiated from carrier gas by sensing the influence of molecular-weight variations or the ionization susceptibility of the gas mixture flowing into the GC detector. The molecular weight of a gas and its thermal conductivity are related; light molecules can diffuse faster between different thermal zones compared to heavier, slower molecules. The thermal conductivity of light molecules is thus higher than that of heavier molecules and directly influences the rate of heat transfer within a gas. Sensing the changes in heat-transfer capacity from a hot, constant-temperature element to a surrounding gas is the basis of the thermal conductivity detector or katharometer. Hydrogen in certain concentrations in hydrocarbon samples is somewhat an exception to this general statement because of problems with TC detector response. Hydrogen has the lowest molecular weight and highest thermal conductivity and thus produces a negative peak at certain concentrations in TC detectors that are intended to operate with heavier-molecular-weight carrier gases. The result is the formation of a "W"-shaped hydrogen peak. This problem can be avoided by using a mixture of hydrogen and helium as the carrier gas when the hydrogen concentration is above the inversion concentration. A second problem is that the hydrogen response is linear to only about 15%; TC detectors must be individually calibrated for higher hydrogen levels.

Figure 9.7 shows a sectioned diagram of a TC detector block with filaments installed in the gas flow passages of the detector. The hot-wire filaments are wound from a material that changes resistance greatly with temperature. An external Wheatstone bridge circuit senses the minute changes in filament temperature caused by changes in the heat-conducting capacity of the surrounding gas. Two flow paths through the detector are

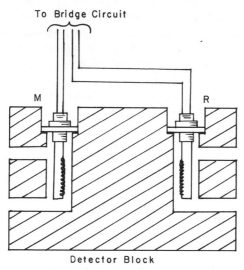

Figure 9.7. Thermal conductivity detector for gas chromatograph (M, measuring hot wire: R, reference hot wire).

shown in Figure 9.1 with one side used for a reference thermal conductivity measurement of the carrier gas and the other side used to sense the combination of carrier gas and eluted components. Precise control of carrier gas flow rates, detector current, and detector block temperature are required for stable operation. Other resistive elements such as thermistors are sometimes used in TC detectors, especially if the sampled gases can deteriorate the hot-wire filaments or if heat-induced catalytic decomposition of the separated components might occur.

The hot-wire filaments of a TC detector are connected to Wheatstone bridge circuits as resistive elements in the legs of the bridge. The bridge imbalance caused by changes in the resistance of the measuring filament during component detection can be amplified and transmitted to recorders or other data conversion circuits. Any electrical circuit for these detectors must keep the current supplied to the hot-wire filaments constant because changes in filament temperature can cause resultant changes in detector sensitivity. Extremes of this situation can be imagined where, in one case, no current was flowing in the filaments so that no heat could be removed to change their resistance. The other extreme case would be one in which the filaments were driven so hotly that noise and baseline (bridge balance) instability from small carrier gas flow disturbances would become excessive. Another distinct possibility is burning out the filaments because of high filament currents.

The amplifier circuit of the TC detector is designed to develop a specific output change for input voltage change of each particular component detected from the bridge circuit. This is called *attenuation*. Filament resistance increases with temperature, so the voltage drop across that filament of the bridge will increase if the current increases. Bridge balance controls can offset any small imbalance caused by minute differences inherent in individual filaments or filament pairs. However, a filament that is many times as resistive as it would otherwise be if operated at a lower temperature will likely develop a much larger bridge imbalance when components impinge on it. This characteristic is not only one way to increase the sensitivity of most thermal conductivity detectors, but is also the basis of operation for circuits called *constant-temperature bridge amplifiers*. These circuits develop output signals proportional to the filament current changes required to keep the detector elements at a constant resistance (thus temperature).

Components that might so unbalance a conventional TC detector because of high concentrations or extreme thermal conductivity effects could possibly cause errors in detection of components with much less imbalance capability. This is because of short-term change in detector filament sensitivity. By keeping the hot wires at a constant specific temperature, their sensitivity remains constant so that they can respond to various components as they are eluted without excessive time being required for reestablishing detector stability. Constant-temperature bridge amplifiers can monitor TC detector bridge current requirements with such sensitivity and speed that the temperature of the components eluted across the filaments does not change sufficiently to cause any practical change in sensitivity of the components. This type of amplifier is useful when very different concentrations of very different thermal conductivity components must be accurately quantitated. Thermal conductivity detectors are considered universal detectors; they respond to all components having a molecular weight different from that of the carrier gas. This can sometimes be a problem if water, air, noble gases, or other components in the sample or carrier gas cause interference in the detection of the components of interest.

Ionization detectors sense the small electrical currents developed by charged particles in sample as it is exposed to an ionization field. These detectors have demonstrated remarkable selectivity and sensitivity. The most common type of ionization detector, the flame ionization detector (FID), senses the extremely small changes in electrical conductivity of a hydrogen flame while a sample is burned. The flame conductivity is influenced by ions and electrons developed during the combustion of the sample. The FID burner tip is connected to one side of a high-impedance, high-voltage source with the collector element of the detector connected to the other side. An electrometer amplifier is used to amplify the extremely small electrical currents carried by the flame while sample components eluting from the column modulate the conductivity of the flame. This description is an oversimplification of flame ionization, but the fundamental principles apply equally well to other ionization detectors even if the ionizing field is developed by a small radioactive source. The electron capture detector and the helium detector use β-radiation sources to ionize the separated components flowing into the detector. The electron capture detector is extremely sensitive to halogenated compounds often found in pesticides but virtually insensitive to aliphatic hydrocarbons, alcohols, and ketones. The helium detector is used for the ultratrace analysis of permanent gases.

Figure 9.8 shows a very simplified diagram of a flame photometric detector (FPD). This detector does not use the flame as a variable conductance circuit element, but rather as a mechanism to excite the sample component into emission of light. The difference between a FID and FPD are predominantly in the collector; in the FID this element is generally a cylindrical electrode located near the top of the flame tip, whereas in the

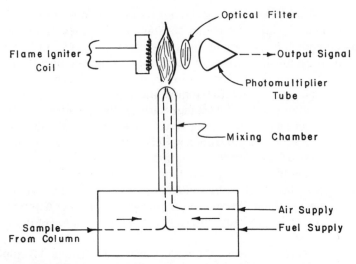

Figure 9.8. Principle of operation of the flame photometric detector. Courtesy of Analytical Instrument Corp.

FPD the collector is an optical filter and photomultiplier tube assembly. The intensity of specific wavelengths of light emitted by certain components as they are burned can be related to their identification and concentration. Flame photometric detectors are extensively used for the trace analysis of sulfur and phosphorous compounds in petroleum products and other chemicals.

Often a price must be paid, however, for the selectivity and sensitivity of some ionization detectors. The trade-offs can be reduced linearly, a requirement for more complicated mixtures of specific carrier gases, susceptibility to contamination from background impurities such as oxygen or water in the sample, and the need for frequent specialized calibration routines. Calibration of process gas chromatographs using packed columns and TC or FID detectors is often more straight forward and somewhat simpler than the routines used for more exotic combinations of columns and detectors. There is much data available to the chromatographer concerning the retention times of various substances and their relative responses to different detectors.[10-13] These data can be effectively used for component identification if all of the variables that affect column and detector performance are rigorously controlled and if the column and detector meet certain requirements relating to resolution and reproducibility of results.

9.2.6. Signal Conditioning

Detector output signals are measured as a function of time with the starting point usually the time of sample inject. The detector output just prior to the start of an analysis should be stable, free of excessive noise, and at the level that the detector will produce while running on pure carrier gas. Output signal traces are usually made with a strip chart recorder or other x–y plotting device showing increasing detector signal on the y axis and analysis duration along the x axis. Although the reference time used to show the start of an analysis cycle is often the event of sample injection for on-line process GCs, other types used primarily in laboratories use the location of the response of a detector to an unretained reference compound introduced during sample injection. The time required for a component to elute from the column and cause a detector response is measured from the analysis cycle starting time reference to the detector peak maxima.

Retention times are a function of the kind of solvent and the amount of it used in the column, the volume flow rate of carrier gas, the column temperature, and the column length. Data on retention times of various standard chemical compounds can be compared to retention time data of unknown peaks if exactly the same column operational features have been rigidly adhered to in both cases. This is also true of data relating to the response of a particular detector to a standard component. These data are valid only if the operating conditions specified during the original determinations are used when the later analytical separations are made. Unfortunately, this is often impossible with on-line process chromatographs because of the need to keep these analyzers in service and dedicated to a specific analysis.

Process GC detectors can be considered to be concentration-measuring devices that provide outputs proportional to the amount of component in the detector at a certain time. The output response a detector develops for a certain amount of one component might be quite different from its response to a similar amount of some other component. Because of this response variation and the fixed nature of the GC installation, external standards are often used to determine the relative response of a particular detector in terms of calibration engineering units.

Figure 9.9 shows a typical chromatogram of the separation of two components where the geometry of each peak approaches the area distribution of an ideal Gaussian curve. The area of this type of peak can be closely approximated by multiplying peak height times peak width a $\frac{1}{2}$ peak height as long as the peaks are symmetrical. The resolution of the

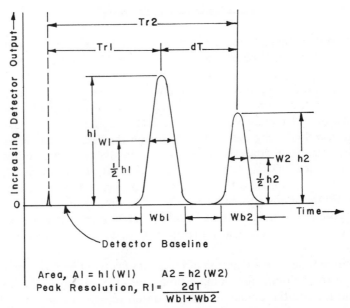

Figure 9.9.   Chromatogram showing the important characteristics of two-component peaks.

peaks; therefore, the efficiency of the separation can also be determined by the ratio of twice the difference between the retention times of peak maxima dT divided by the sum of the peak's baseline widths, W_{b1} and W_{b2}. This ratio should never be less than 1.0 and should typically be 1.5 or higher to achieve 99% resolution. The column must separate the components sufficiently that the detector can generate the ideal Gaussian peaks required for accurate determination of peak areas.

    The conversion of detector signals to engineering units representing the various concentrations of components in the sample can be accomplished by several methods. Peak height measurements are often used after absolute calibration against an external standard. This method is convenient and reasonably accurate if the response factor of the detector to the standard is the same as its response to the measured components. The minimize discreptancies, the standards to be used for calibration should be the same as the components being analyzed and should be injected in quantities representing 80 to 100% of their maximum calibrate range. As long as chromatographic conditions are stable, the height of a particular peak can be scaled to the height of calibration peak of the same component. The ratio of the peak heights is multiplied by the amount of that component contained in the calibration standard, thus converting that peak height to the same engineering units used for the calibration standard.

Peak heights are often subject to changes stemming from problems with sample injection or column operation. Peak areas, on the other hand, can accurately represent the weight percent of each component if the detector response to each component is considered. FID relative sensitivity and TC detector molecular weight factors[10] can be used to compensate various component peak areas relevant to the peak area of a standard component such as benzene. If all components in the sample are separated and measured, normalization of the corrected peak areas is performed by dividing each corrected peak area by the sum of those areas and then multiplying this ratio by 100. The result is in component weight percent. Digital computers have enhanced chromatogram interpretation by utilizing more sophisticated mathematical techniques for accurately determining areas of asymmetrical peaks, correcting drifting baseline references, and allowing specific calibration factors to be stored in computer memory for conversion of peak areas to engineering units.

The control and computational programs used in most digitally controlled chromatographs can often perform the GC monitoring, valve control, detector data conversion, and communications interface functions for more than one analyzer. These programs are also capable of scheduling automatic calibration cycles, running diagnostic routines, and performing special calculations. Technological advances and microminiaturization might eventually allow all chromatograph features and control functions to be incorporated into something on the order of the present-day microprocessor integrated circuit.

9.3. LIQUID CHROMATOGRAPHS

On-line analysis of process samples by gas chromatography is practical when the sample can be injected into the analytical column(s) as a vapor. Samples of thermally sensitive, high-molecular-weight polymers, nonvolatile compounds, inorganic salts, or complex biochemicals are seldom suited for GC analysis. High-performance liquid chromatographs (HPCL's) are finding increasing use for the analysis of these materials. HPLC analyzers are primarily used now in the laboratory and are not intended for continuous, single-analysis operation. However, this will probably change as applications are defined and HPLC systems perfected for use in continuous, on-line service.[14]

Figure 9.10 is a block diagram of an HPLC. The solvent pumping and flow control system provides a mobile phase that moves the sample

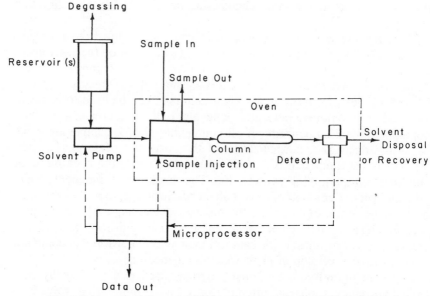

Figure 9.10. Block diagram of high-performance liquid chromatograph.

through an analytical column. Most HPLC detectors are designed to measure liquid stream parameters such as refractive index (RI) or ultraviolet (UV) absorbance; however, electrochemical types are also available. Optical coupled circuitry and optical gates may speed development of micro-HPLC systems employing fluorescence, RI, or UV absorbance detectors in combination. The development of microprocessor-controlled, high-pressure solvent pumps has been a major factor in establishing reproducibility in HPLC systems. Solvent flow control in especially important in gradient-elution HPLC systems where the solvent composition is changed during the analysis. This is somewhat similar to temperature programming in GLC where components are separated by fundamentally changing the column operation during one cycle.

One unique feature of HPLC mobile phases is that they can sometimes be an active agent in the separation process. For example, a pure water mobile phase can be modified to change ionic strength with regard to certain components in the sample and thus influence that separation. Nonaqueous mobile phases require purification, recycling, and degassing systems for continuous HPLC operation that have not yet been perfected for most field applications. The HPLC sample injection system is similar to systems used in GC's but built to withstand much higher pressures and use smaller sample volumes. Liquid chromatographs are being increasingly used in laboratory analysis primarily because of improved column packing techniques and better data reduction, instrument control, and diagnostic routines available with microprocessor based systems.

HPLC columns are very efficient and achieve this over a short length when compared to GC columns. Extremely small particles of silica only 6 to 10 μm in diameter are densely packed under high pressure into a column typically 4.6 mm in inner diameter and 15 to 25 cm long. This type of technology has allowed manufacturers to provide performance guarantees achieved by rigid quality control and automated testing systems.[15,16] Spherical glass beads of 30-μm diameter coated with much smaller silica beads (2000 Å diameter) are sometimes used as the support for bonded-phase columns that use partitioning or adsorption mechanisms to achieve separation. Porous particles such as silica or cross-linked polystyrene can use size exclusion mechanisms or, if the packing has charge bearing functional groups on its surface, ion-exchange separation mechanisms. Much more detailed information is readily available to those readers applying various HPLC column configurations to analysis problems.[17–19]

HPLC detectors typically operate using optical absorbance principles in the IR, visible, and UV regions of the electromagnetic spectrum. Both fixed- and variable-wavelength detectors are available depending on analysis requirements. Refractive Index and UV fluorescence detectors are used in applications requiring much higher detector sensitivity. Electrochemical or polarographic detectors can have sensitivities in the picomole (nanogram) range and are highly specific to the detection of certain compounds in complex matrices such as body fluids or natural products. Flame ionization and thermal or heat adsorption detectors have been applied to HPLC analyzers, but their use requires more complicated instrumentation, and data interpretation is seldom as straightforward or well documented as is the case with fixed-wavelength UV absorbance detectors operating at 254 nm. By combining UV and RI detectors in one analyzer, an almost universal detection capability can be obtained.

HPLC has been included in this chapter because this rapidly developing technology seems to be ideally suited for microminiaturization; in fact, the very small internal-diameter tubing, sample volumes, and low solvent flow rates that are often typical of HPLC systems may someday be adapted to chromatographs in the form of monolithic circuits or single-chip analyzers. This has already been accomplished by using GLC techniques,[20] and HPLC microanalyzers may soon follow this example.

9.4. GENERAL COMMENTS

On-line process gas chromatographs are extremely useful instruments for multistream or multicomponent analyses. For most applications, accuracies of ±1% and reproducibilities of ±0.5% are achievable. Although initial costs are somewhat high, the GC may be an economical choice when compared to other multiple, dedicated, on-line analyzer systems. Microprocessor-controlled GC's are being increasingly used for real-time stream composition measurements in complex process control applications.

CHAPTER

10

MAINTENANCE AND TROUBLESHOOTING

Despite its title, this chapter is divided into four equally important sections entitled "Training," "Maintenance," "Troubleshooting," and "Documentation and records." These four sections are closely related, dependent on each other, and necessary for the implementation of a well-planned, effective maintenance program.

Small plants with a limited workforce can still apply the basic principles presented here to achieve a highly successful maintenance program. For example, formal classroom training might be replaced with a series of one-on-one instruction periods covering instrument design or a specific plant application.

A successful maintenance program is also a flexible one; there should be plenty of opportunity for maintenance technicians to input information into the system and participate in writing operating and calibrating procedures, spare parts setup sheets, and log sheets. Technician participation often results in the development of simplified procedures without omitting essential details or taking unwarranted shortcuts.

10.1. TRAINING

The amount of training required to maintain process analyzer systems depends on the maintenance concept adopted. An "operate until failure" concept is totally unacceptable with on-line analyzer systems. In most cases some sort of routine maintenance or preventive maintenance program would seem the best. Will the analyzer systems be maintained by plant maintenance personnel or an outside contractor? Will repairs be limited to a printed circuit replacement philosophy, or will component level repairs be attempted? In this chapter it is assumed that maintenance will be performed by plant maintenance personnel and that repairs will be limited to the replacement of printed circuit boards and major sample system components. A large electronics shop with skilled repair personnel is not required in this case.

Informal training can usually be obtained when new plants or new systems are started up and when experienced technicians are available for training new technicians in a particular maintenance or operating area. This "experienced technician—rookie technician" training concept has one minor disadvantage; some training information is lost each time a new rookie is trained. For this reason, periodic refresher training sessions will probably be required even for some so-called experienced technicians. Some sort of formal classroom training is highly desirable if it can be implemented economically. A number of preprepared topical handouts are usually given to technicians during formal training; these handouts become the property of the technicians and are intended to provide background reference material. Quizzes may be given during formal training, but problem-solving exercises may prove to be a more effective training aid.

10.1.1. Startup Training

New plant or process startups provide excellent training opportunities for technicians who will be expected to maintain the analyzer systems once they are implemented. Generally an experienced engineer, analyzer specialist, or vendor's service representative is responsible for the startup and initial operation of new analyzer systems. Inexperienced technicians can learn a great deal by working with and observing "the expert" during this phase while he is troubleshooting the system. Technicians should ask questions when they do not understand what the expert is doing or why he is doing it. There must be good two-way communication at this time for the startup training to be effective. The supervisor who requests or makes arrangements for the services of the analyzer expert should obtain the consent of the expert in advance when analyzer training is a primary goal.

The analyzer expert checks the sample system design and verifies that the correct materials of construction have been used and makes modifications as necessary to the sample system and analytical equipment to assure that everything functions as designed or as the owner intended. The analyzer expert calibrates the analyzer when the system is operational and evaluates its performance preferably over several days because design flaws tend to surface during the first few days of operation under process conditions. When this happens, the analyzer expert can consult with the factory and make repairs on the spot. The operating department that wants to use the analyzer should not pressure the analyzer expert to put the system on-line prematurely. Sufficient time should be allowed for the manufacturer's response to problems found during the evaluation period and implementation of necessary field repairs or modifications. It is during this testing and evaluation period that individual attention can be given to technicians receiving analyzer training.

An evaluation report should be written at the end of this formal evaluation period. The following topics should be included in the report when applicable:

1. Condition of equipment as initially observed, including notes on obvious shipping damages.
2. Required hardware modifications if the equipment is not immediately capable of being installed or commissioned.
3. Required software or programming modifications.
4. Corrected data sheets and "as built" drawings with an up-to-date spare parts list.
5. Tentative operating and calibrating procedures.
6. Recommendations for proposed long-term improvements and required analyzer maintenance.

Copies of the evaluation report should be kept with the analyzer instruction manuals and placed in area maintenance files for future reference.

10.1.2. On-the-Job Training

Continuous on-the-job training is required after plant or process startup because of technician rotation, vacations, and terminations. Area maintenance technicians should accumulate valuable experience with on-line analyzer systems and begin to establish maintenance history files during the first 6 months of operation. Some technicians prefer compiling information books containing copies of instruction manuals, evaluation reports, and instrument data sheets. Unfortunately, these technicians usually cannot remain on analyzer work indefinitely; therefore, they are often asked to train new technicians in analyzer repair and maintenance procedures. The effectiveness of this type of on-the-job training depends on the skills and abilities of the experienced technicians and the willingness to learn of the inexperienced technicians.

On-the-job training may become less effective as the plant grows older because of poor motivation or lack of good communications between experienced and inexperienced technicians. The reasons for following certain maintenance procedures may become obscured with time and result in the adoption of less effective procedures.

Additional on-the-job training can be obtained if the vendor's service representatives must be called in to upgrade the system or make analyzer repairs. If a severe system failure should occur that extends beyond the routine maintenance capabilities of the plant, technicians working in the area should be scheduled to assist the service representative as required.

10.1.3. Semiformal Classroom Training

Serious consideration should be given to instituting periodic semiformal classroom training if the analyzer maintenance program begins to fall short of expected goals. The training course should be limited to coverage of process analyzer systems actually used on the plant and should be fairly well balanced between classroom instruction periods and field trips where technicians can get "hands on" experience. The class size should be limited to 8 to 10 persons to prevent overcrowding in the field.

Classroom lectures should be based on material contained in specially prepared manuals given to each participating technician. Suggested topics for the analyzer training manual might include:

1. Theory of operation.
2. Routine operation.
3. Calibration techniques.
4. Problems and solutions.
5. Maintenance aids.

The same topics can be covered with each type of analyzer system discussed. The "Theory of Operation" and "Calibration Techniques" sections are intended to familiarize new technicians with various analyzers such as pH meters, corrosion monitors, ultraviolet photometers, and so on.

The "Routine Operation" and "Problems and Solutions" sections are intended to inform the technicians of actual plant problems that have been encountered and subsequently solved.

Area tours are used to inspect operating analyzer systems. These field tours can provide an excellent opportunity to demonstrate abstract training pertaining to the effects of corrosion on sample systems or more concrete ideas such as the effectiveness of local pressure gauge and flowmeter monitoring. Technicians can experience "hands on" training during the calibration of operating analytical systems. Minor repairs or cleaning can be handled by the analyzer training class as they investigate these systems.

Some items that might be covered under "Maintenance Aids" include:

1. Spare parts list.
2. Log sheets.
3. Instruction manuals.

4. Vendor list.
5. Operating and calibrating procedures with calibration data sheets.

Training manuals are generally limited in size, so usually only samples of data sheets, manuals, log sheets, or procedures are included. The purpose of this section of the training manual is to inform technicians about the various maintenance aids available and where these aids can be obtained.

Semiformal classroom training is an effective method of providing the necessary specialized information and is often quite popular with analyzer technicians. Armed with new ideas and their own personal copy of the training manual, technicians usually look forward to their next field assignment where they can apply their newly acquired knowledge.

10.2. MAINTENANCE

The job of keeping on-line analyzers and their sample systems operational is the responsibility of the analyzer maintenance group. One of the most effective concepts for analyzer system maintenance involves formation of an analyzer maintenance group consisting of four or five analyzer technicians reporting to an instrument engineer whose specialty is on-line analyzer systems. The instrument engineer is responsible for:

1. The specification, design, installation, checkout, and evaluation of new analyzer systems.
2. The coordination and scheduling of routine maintenance on existing analyzer systems.
3. The personal investigation of nonroutine trouble calls and major analyzer system failures.
4. The initial training of analyzer maintenance technicians and subsequent training of new technicians rotated into the group as replacements.

Another popular technique used in analyzer maintenance programs involves basic analyzer training for all instrument maintenance technicians and advanced analyzer training for instrument technicians working primarily with process analyzer systems. The duties of the plant instrument engineer responsible for analyzer systems remain the same.

Responsibilities of the analyzer maintenance technicians include making daily inspections of analyzer sample systems and performing routine weekly calibration checks. The purpose of the daily inspections is to check sample system flows, pressures, and temperatures and verify that the analyzer output appears normal. If an abnormal condition is noted, it should be corrected at that time with a description of the problem and the corrective procedure listed in the maintenance history records of the analyzer. Any inexperienced maintenance technicians should accompany the experienced technician during these inspections. The inexperienced technicians should also participate in routine weekly analyzer calibration. Zero and span settings should be logged along with data on sample system flows, pressures, and temperatures. The weekly log provides information on the stability of the on-line analyzer and its sample system.

The analyzer program may be able to be handled by one engineer and one or two maintenance technicians in small installations. Contract maintenance may be preferred on large, complex instrument systems such as the main plant computer system or where a large chromatographic installation has been provided. However, substantial amounts of money can still usually be saved by training local plant personnel to perform routine maintenance services.

Regardless of the type of maintenance program selected, the following checklist of items should be considered:

1. Who has the prime responsibility for the overall maintenance program?
2. Will in-house training be provided, and to what extent?
3. Does the design provide for adequate clearances from a safety and ease of maintenance standpoint?
4. What type of sample tap will be provided?
5. How much sample conditioning is required for the sensor?
6. Should the process analyzer system be housed in an analyzer house?
7. Is pressure, flow, and temperature monitoring adequate?
8. Is the analyzer system in closed-loop or interlock control?
9. Does the design need pressure-relief devices or check valves?
10. Are purge and/or calibration gas facilities required?
11. How is the sample being disposed of?
12. Would daily, biweekly, or weekly checks be provided?
13. Are logs, instruction manuals, historical records, or trouble call files needed? Where will these be set up?
14. Will an instrument repair facility complete with test equipment be needed to support the maintenance effort?

10.2.1. Personnel

Technicians showing an interest in obtaining the necessary training for instrument and electrical maintenance are more likely to seek to expand their job skills by studying process analyzer systems. On-line analyzers are often considered a separate phase of the instrument maintenance program of the plant. Technicians who have expressed an interest in process analyzer systems should be given a chance to participate in the analyzer maintenance program; people who are happy with their work tend to do a better job.

Analyzer maintenance technicians must have adequate technical backup; a well-organized maintenance file room with copies of instruction manuals, vendor catalogs, analyzer purchase orders, and as-built drawings are all essential because of the complexity and sophistication of most analytical instrumentation.

One of the most overlooked areas of analyzer maintenance support is that of facilities in which to efficiently carry out this work. The type of work space required will depend on the extent of the applications of analyzers throughout the plant and the maintenance philosophy that has been selected by plant management. Large operating areas using many analyzers may be better suited to an area maintenance concept as opposed to a centralized facility. Much time can be lost while technicians attempt to find effective transportation for delicate test equipment and system components. Even if no loss or damage occurs because of excessive handling, there is a possibility of environmental damage to some sensitive analyzer components from rain, dust, or sometimes direct exposure to sunlight. Frequently the nature of analyzer maintenance is not compatible with the efforts of other maintenance groups. This has been shown to be the case whenever a pipe welding or machine shop facility is provided with no additional environmental protection from dust, excessive humidity, or high noise levels. Many corporate managers overlook the utilities usually required for analyzer system maintenance and consequently do not provide the necessary initial capital expenditures to include instrument air, steam, potable water, vacuum or vent systems, oily water sewers, and hazardous chemical storage facilities that almost any modern analyzer maintenance facility now requires.

Analyzer maintenance groups must often perform additional highly specialized repairs to digital computers or other complex electronic subsystems used for controlling chromatographs, tank car weighing systems, or remote telemetering equipment. The particular requirements of each maintenance assignment must be comprehensively reviewed to ensure that any specialized test equipment or repair tools are available when needed.

10.2.2. Materials

Spare parts availability, technical obsolescence, and equipment warranties are major areas of concern in analyzer maintenance programs. What is the availability and delivery of a replacement analyzer, critical spare analyzer parts, or critical sample system parts? Should a good compliment of parts be stocked, or only critical parts? Is there a second supplier? Can parts be substituted? These are some of the questions that should be asked by personnel responsible for maintaining on-line analyzer systems.

Recent material shortages have shown that the delivery of exotic materials such as tantalum, Inconel®, and Hastelloy® can exceed 6 months to a year. Extreme care must be exercised to ensure that everything required for the total analyzer installation is ordered with enough lead time to offset long delivery schedules. Less exotic materials with higher corrosion rates can sometimes be substituted until more exotic materials arrive, but it is extremely important that temporary designs and substitute parts do not become permanent fixtures in the final installation.

There may be a delay of up to a year or more from the conception of a major project until its final installation. Instruments and analyzers may be out of warranty before they are installed and tested. A manufacturer's 90-day warranty is of little value if the warranty period begins when the part is sold or delivered. Currently there is a trend for instrument manufacturers to offer the user a "one year from date of installation" warranty. This appears to be a much more reasonable warranty from the viewpoint of the industrial user. In some cases instrument warranties of up to 5 years may be offered. These longer warranties reflect the manufacturer's confidence in his product and a general feeling that the instruments and analyzers should be rugged and reliable enough to operate free of failures for more than just a few months.

Analyzer obsolescence usually varies from 5 to 15 years; this is due in part to rapidly developing technology and in part to general wear and tear. The development of microprocessors and user prompting or menu software is a prime example. The 16-bit machine has made its way into the marketplace; when will 32-bit or 64-bit systems be incorporated into process analyzer systems? An on-line analyzer that has been in more or less continuous service for 10 years can be considered exceptional.

10.2.3. Design

Equipment should be easy to maintain, and this premise should be kept in mind when the physical arrangement of analyzers and their sample system layout is being planned. Consideration should also be given at this time to including a maintenance engineer on the design team.

The following topics should be considered in order to simplify the maintenance effort:

1. Safe startup and shutdown of equipment.
2. Accessibility and ease of replacement of system components.
3. The use of industry standardized brackets and mounting hardware.
4. The use of a minimum number of components versus installing dual systems.
5. Any special requirements for purge, cleaning, or calibration.

During the course of normal plant operations analyzer sample systems often must be started up and shut down numerous times, so they should be easy to isolate from the process. It should be kept in mind that some sample systems must be frequently drained, purged, vented, cleaned, and flushed, so adequate valving should be provided for these operations. Some thought should be given to how trapped samples or cleaning and flushing materials will be later removed from the system. Sample system lines should be sloped so that natural drainage occurs, and drain valves should be provided at low points. All valves should be within easy reach and plainly marked according to their function as sample block, bleed, vent, drain, or purge. Startup and shutdown of the sample system should not cause a "water hammer" effect in liquid systems or a sudden pressure surge in gas or vapor systems to minimize the possibility of analyzer component damage.

System components should be laid out so that they are readily accessible to facilitate efficient removal, repair, and replacement. It is relatively easy to heat trace and lag sample lines when they are initially installed. How easy is it to remove sample system components and reinsulate them when they are replaced? Proper operation of heat-traced sample systems require that hot and cold spots be avoided along the entire length of the sample line. Mounting brackets and sample system hardware should not interfere with heat tracing.

Sample system costs can be minimized by using standard mounting components such as common hardware brackets, Unistrut® channels, and

$\frac{1}{4}$-in.-thick aluminum plate. The hardware brackets and mounting plate make it easy to mount components and provide a layout that is easy to maintain because major system components are replaced in the same positions at drilled and tapped holes in the plate. The aluminum plate is easy to drill, saw, or punch, thus minimizing initial installation time. Tapped holes should be used for component mounting to avoid hidden fasteners. The mounting plate and components should also be easy to remove from field cabinets or analyzer racks.

Design engineers try to minimize the number of sample system components in an effort to increase the system reliability by reducing the total number of parts required. However, when dealing with dirty streams or critical applications, it may be desirable to install dual components. Complete dual on-line analyzer systems may be required in critical applications where downtime cannot be tolerated.

10.3. TROUBLESHOOTING

There are a number of techniques used in troubleshooting analyzer systems. Equipment operated until failure might be relatively easy to troubleshoot but may also be more difficult to repair effectively. Furthermore, this technique is generally unacceptable for use with process analyzer systems because the user wants the best possible qualitative and quantitative information over the longest period of time. A preventive maintenance program most closely achieves this goal because periodic equipment inspections are regularly scheduled; any routine or special maintenance work can be performed at that time if necessary. Unexpected total system failures can usually be minimized after the inspection cycle period has been properly established.

Another system troubleshooting technique involves inspection and servicing of equipment only at the request of the operating department. Semiformal work requests may be written to recalibrate an analyzer or replace steam tracing on a sample system. The requests are briefly written, but they do inform the analyzer maintenance technician that something is wrong with the analyzer's sample system or hardware. Work request information should not be taken literally because "recalibrate the analyzer" may really mean "unplug the sample system" after a careful investigation of the problem has been made.

A trouble call results when a sudden analyzer system failure has been observed. Sudden failures can occur in spite of the fact that a good preventive or routine maintenance program has been instituted. A broken rotameter, blown fuse, or other similar hardware failure would all result in placement of a trouble call to the instrument department. Priority is generally given to trouble calls because the user needs the system placed back into service as soon as possible. Response to trouble calls is referred to as fire fighting because repairs must be made as quickly as possible so that attention can be given again to the routine analyzer work load.

10.3.1. Preventive Maintenance

It is easy to say that every plant should have a good preventive maintenance program, but it is somewhat more difficult to establish one. There is a genuine concern in that a regularly scheduled preventive maintenance program requires a definite allocation of workforce and time regardless of whether any maintenance repair work is required. There is also the question of plant size; how large a plant or how many analyzers must the plant have in order to justify establishing a preventive maintenance program? There is no clear-cut answer to this question. A preventive maintenance program might be justified with half a dozen complex analyzer systems using GC's and automatic titrators or a large number of less complex analyzers such as pH meters or corrosion monitors.

Some work that might be covered under a preventive maintenance program includes:

1. Checking and adjusting valve packing and stroke length of throttling valves.
2. Checking the cleanliness of glass tube rotameters and optical cells.
3. Lubricating mechanical and electromechanical parts.
4. Adding reagents to wet chemistry analyzer sample systems.
5. Checking and adjusting sample system flows, pressures, and temperatures.
6. Checking analog panel meters for cracks, mechanical zero shifts, and sticking meter movements.
7. Checking electrical potentiometers for smoothness of operation and possible short circuits or open circuits.
8. Physically cleaning the equipment by removing corrosion products or built-up dirt deposits.

Dozens of items similar to the above can be added to a preventive maintenance checklist.

The purpose of the preventive maintenance program is to provide a minor tune-up and overhaul of the process analyzer system before major repairs are needed. Annual maintenance costs can usually be minimized by keeping analyzer systems in good operating condition. The need for a preventive maintenance program increases with the age of the equipment just as the likelihood of a sudden, catastrophic equipment failure also increases with age.

10.3.2. Routine Inspections

Routine inspections are made in a preventive maintenance program but do not constitute the preventive maintenance program in themselves. For example, control valve packing leaks would probably be found during a routine inspection, but valve stroke length problems or defective instrument potentiometers might not be detected. Critical flows, pressures, and temperatures should also be checked and adjusted if necessary during a routine inspection.

Weekly inspections of process analyzer systems are usually made at plants that have adopted total in-house maintenance program responsibility. Analyzer technicians typically establish a habit of making routine equipment inspections the first thing every morning with a survey of carrier gas supplies, reagent flows, and system pressures the last thing every afternoon, or at some time mutually agreeable with the operating and maintenance area supervision. The period between routine maintenance inspections might vary depending on the number and complexity of the analyzer systems; a consensus of manufacturer's recommendations and field experience is usually followed. Some analyzer experts prefer biweekly inspections until sufficient maintenance data are gathered to justify a change in the inspection period.

Routine inspections do not involve the issuance of a maintenance checklist, computer card, or other aid to remind the instrument technician to perform certain specified maintenance tasks. Only work that needs to be done is done at that time; the routine inspection program results in a routine maintenance program. Routine maintenance programs are popular because of their lower costs on a day-to-day basis; the preventive maintenance costs are not incurred. However, annual maintenance costs may actually be higher than those of a preventive maintenance program because of system failures and downtime.

Large plants may also conduct annual plant analyzer audits. Ideally the audit would consist of a plant tour and inspection of every on-line analyzer system by a team made up of an analyzer expert, a representative from the operating area using the analyzers, and the analyzer technician responsible for maintaining the analyzers. The audit team carefully inspects each analyzer and sample system and notes any items needing repair or restoration. A formal audit report is issued to members of management after the tour and inspection is completed. Some of the items that should be reported in the audit include:

1. Analyzer tag number or other equipment identification.
2. Make and model number of the analyzer.
3. Brief description of the application.
4. Data and initials of the auditing team.
5. Notes on "as found" condition of the equipment.
6. Recommendations for improvements.

The analyzer audit report informs plant management of the effectiveness of the process analyzer maintenance program and tends to generate additional analyzer restoration work.

10.3.3. Trouble Calls

The operating department usually originates a trouble call whenever an analyzer output transmission loop fails to respond in a predictable manner. An analyzer technician is assigned to investigate the problem and make the necessary repairs and, to do this efficiently, must systematically check signal levels at various places in the signal loop. A simplified analyzer loop diagram is shown in Figure 10.1. A logical place to begin troubleshooting the system is at the control room termination block, where the 4- to 20-mA DC analog current signal is converted to a 1- to 5-V DC analog voltage signal by passing that current through a 250-Ω resistor. A comparison of loop current, control room terminal voltage, and panel instrument or computer data display is given in Table 10.1.

It should be noted that the analyzer signal loop shown in Figure 10.1 utilizes a live zero or an output greater than 0 mA DC or 0 V DC even when there is no on-scale analyzer input. Therefore, there should always be at least 1 V DC on the field termination block terminals. If the field termination voltage is less than 1 V DC, there is probably a short circuit in the analyzer output or a component failure in the output circuit of the analyzer. The wires to the panel instruments and computer can be lifted one at a time to isolate a faulty device inside the control room.

If a problem has been located in the field instruments or equipment, a careful check should first be made of the sample system of the analyzer. Well over 75% of all analyzer problems are sample system problems. Typical sample system problems include:

1. Air leaks into or process leaks out of the sample system.

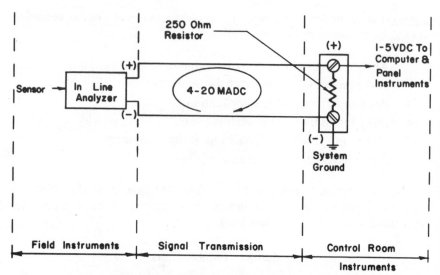

Figure 10.1.   Simplified analyzer loop diagram.

2. Plugged filters, valves, orifices, sample lines, check valves, or outlet lines.
3. Dirt or film buildups on optical cell windows.
4. The accumulation of moisture or process condensate in the sample system.
5. Incorrect sample system flows, pressures, or temperatures.
6. Insufficient amounts of reagents, buffers, or other chemical agents in a wet chemistry system.

The sample system of the analyzer should be thoroughly checked out before a detailed check is made of the electronics system. If the trouble

Table 10.1.   Typical Instrument Outputs

Analyzer Output Current, mA DC	Field Terminal Voltage, V DC	Panel Instrument or Computer Data Display, %
4	1.0	0
8	2.0	25
12	3.0	50
16	4.0	75
20	5.0	100

appears to be in the electronics unit, a few simple initial checks should be made as follows:

1. A visual inspection should be made of power cords, fuses, and pilot lights.
2. UV and IR lamps should be burning bright and steady in photometric analyzer systems.
3. A visual inspection should be made to detect overheating, broken circuits, loose connections, or corroded contacts.

10.4. DOCUMENTATION AND RECORDS

The amount of documentation required to support an effective analyzer maintenance program depends on the size of the plant, the number of analyzer systems, and the size of the trained technician group responsible for repair and maintenance of those systems. Record keeping at large plants sometimes seems burdensome; in most cases it is necessary to assure a smooth, continuous operation. Three types of documentation and records commonly used are maintenance logs, status reports, and posted information. Maintenance logs are kept for the benefit of the maintenance department as a historical record, status reports document work and inform management as to what was done, and posted information helps keep instrument technicians informed of the latest information and changes.

10.4.1. Maintenance Logs

Maintenance logs are intended to give a service history of pumps, compressors, instrument valves, analyzers, and other equipment. Card files that carry identification information on each instrument along with a calibration and repair history can be used to inform the technicians about the type of equipment they can expect to find in the field.

Another type of instrument log is used in preventive maintenance programs where a central computer system prints a card for each analyzer system that is due for periodic service. This card also contains general information that identifies the instrument and specifies the maintenance procedures to be used. Preventive maintenance cards usually provide spaces for the maintenance supervisor to acknowledge that the requested work has been done. These cards are then fed back into the computer, and a computer printout can be obtained at any time that shows the status of the preventive maintenance program.

Shift supervisors and first-line operating or maintenance supervisors may also keep handwritten daily logs identifying systems their crews have worked on, describing completed work, and noting any remaining work. Any unusual maintenance problems are usually described in detail and reviewed with all maintenance technicians. These handwritten logs are usually contained in large cloth-bound notebooks and are extremely useful in providing background information when a change is made in supervision or when engineers want to investigate something that has happened on a night shift or during weekends and holiday periods.

Another type of log is kept in close proximity to each installed analyzer system. These logs consist of standard forms that are filled in whenever routine or special analyzer inspections are made. This inspection log usually consists of two main information columns: one labeled "as found" and the other labeled "as left." Whenever settings for the analyzer are adjusted, they should be noted in the local log book, items such as analyzer zero and span, power supply voltage, calibration standard reading, sample flow and temperature, or sensor pressure are examples. This log is especially useful to instrument technicians servicing these systems because gradual changes in a zero or span pot setting might indicate a buildup of dirt in the system or electronic aging. Drifts or settings that change in both directions might be indicative of daytime–nighttime temperature instability of electronic components.

10.4.2. Status Reports

Status reports can be written any time during the life of an analyzer system and are usually intended to bring management up to date on the current status of a particular analyzer system. Analyzer systems are usually commissioned and evaluated while operating for a brief period of time before being turned over to the operating department. The purpose of this evaluation period is to learn about the characteristics of the system such as its accuracy, repeatability, and stability under actual field conditions. Component failures caused by manufacturing defects usually occur during this period and can be corrected prior to the use of the equipment by the operating department. After the analyzer system has been tested and found to be acceptable for routine production use, a status report should be written to summarize evaluation findings and officially turn the equipment over to the operating department.

There are times when the operating department changes process conditions, resulting in the need to recalibrate a process analyzer system. A status report should be generated to cover any necessary recalibration details and inform users of these changes. Status reports may also be written when chemicals, fumes, fires, corrosion, or weather has severely damaged a piece of equipment. The purpose of a status report at this time might be to assure management that the matter has been investigated and should note any recommendations to prevent a future recurrence. This report should give some background information, present a description of the initial condition of the equipment, note how the equipment was left, and briefly discuss various recommendations.

After an analyzer system has been operating for a satisfactory period of time, a final evaluation report or concluding memorandum should be written. This type of status report generally notes the analyzer's service record and informs management that the analyzer expert will no longer be providing close follow-up on the system. In effect, this report states that the system is working satisfactorily and no longer requires close attention from the analyzer expert. However, the analyzer expert may write the final evaluation memorandum on the system one week and then write another status report on it the next week should a system failure occur.

10.4.3. Posted Information

Specific information on individual analyzers should be posted at each analyzer system location. The manufacturer's data sheets, a sample system diagram, and the latest operating and calibrating procedures should be included in this posting.

The manufacturer's data sheets should include the name model and serial number of the analyzer, sample stream flow, and pressure and temperature data and contain all factory calibration data. One purpose of the manufacturer's data sheet is to provide information on critical analyzer parts so that exact replacements can be purchased when needed. Often the manufacturer's original calibration data must be used to confirm that shifts in detector sensitivity have not occurred over a specified period of time. New calibration data sheets showing the latest revision date should be posted if the analyzer has been recalibrated for a new instrument range or application.

Sample system diagrams should be shown on one 8½- by 11-in. sketch when possible. The sketch should show critical valves and specify their normal operating positions (opened, closed, or throttled) so that technicians can tell normal flow path directions at a glance. Valves shown in the diagram should also be labeled with descriptive titles such as block, bleed, vent, purge, laboratory sample, or calibrate in order to increase understanding of how the sample system is intended to work. It is not essential that a parts list be included as part of the sample system sketch; however, the sketch should compliment the posted operating and calibrating procedure.

The posted operating and calibrating procedure should also be limited to a single 8½- by 11-in. page whenever possible; longer procedures may be required with complex instruments such as automatic titrators or on-line GCs. A typical calibration procedure for a photometric hydrogen in chlorine analyzer is as follows:

1. Disable the cycle timer by placing its power switch in the off position.
2. Rotate the cycle timer dial to 25% of full scale.
3. Loosen the optical head clamp screw and pull the head away from the sample cell until the support slide reaches the mechanical stop.
4. Move the analyzer digital display probe from pin 2 on the V/I converter printed circuit board to pin 1 on the output signal printed circuit board.
5. Manually adjust the zero pot until an indication of zero volts is shown on the digital display.
6. Insert the quartz calibration standard into the light path between the light source and the detector on the optical head.
7. Adjust the span control until the digital display indicates the value shown on the posted calibration sheet.
8. After setting zero and span, push the optical head back into its normal operating position and clamp it into position.
9. Readjust the zero control for zero volts on the digital display.
10. Move the digital display probe back to pin 2 of the V/I converter printed circuit board.
11. Start the cycle timer by turning its power switch on.
12. Check the sample system rotameters and the cell back-pressure gauge; readjust controls if necessary to obtain the following:

$$Sample\ in = 1200\ cm^3/min$$
$$Cl_2\ addition = 1200\ cm^3/min$$
$$Sample\ out = 500\ cm^3/min$$
$$Back\ pressure = 5\ psig$$

The preceding procedure was written by an area analyzer technician and illustrates the procedures do not have to be elaborate to be effective.

THE FUTURE

In a few short years the memory capacity of computers will equal and exceed that of humans. Mainframe or host computers will become more like programmable controllers because of the user's desire to add more analog inputs and outputs with real-time processing; programmable controllers will become more like large-frame computers because of the user's desire for increased versatility, which requires more memory. Analog inputs require A/D converters that chew up memory and reduce scan time. The high-speed handling of hundreds of analog inputs and outputs will be required for total plant or laboratory automation with or without robotics. In 1984 32-bit processors will be introduced as a result of this never-ending quest for higher-speed machines.

Growth will also continue in the fields of computer-aided design (CAD), computer-aided engineering (CAE), computer-aided instruction (CAI), and management information systems (MIS). These programs can save thousands of hours of management, engineering, or student time, but they represent off-line computer applications and should not be confused with real-time signal processing or integrated plant automation.

The development of a universal data highway is of prime importance if total process or plant automation is to be achieved. Manufacturers will be forced by their customers to build equipment suitable for use with universal data highways. Some manufacturers are currently incorporating fiber optics in subsystem data highways. Coax cable data highways will probably be replaced with fiber-optics-type universal data highways because of their lighter weight, noncorrosive features, greater signal capacities, and freedom from radio frequency interference (RFI) and electromagnetic interference (EMI).

"Smart" analyzers will become smarter with increased use of artificial intelligence (AI). Laboratory analyzers incorporating AI can now perform multiple correlations and use test results to change input variables to achieve performance goals that the user personally may not recognize on the basis of original program input data. Artificial intelligence will probably get its biggest boost from the defense industry and educational institutions where CAI will provide one-on-one training and education.

The concept of user-friendliness will continue to be of utmost concern in the process control industry. Analyzers incorporating on-board computers will have to be easy to program and use. Simple commands, user

255

prompting, and color CRTs with logical, easily legible displays will continue to be in demand. One manufacturer now offers a gas detector system that speaks when an alarm is acknowledged. In its own unique two-word vocabulary the instrument informs the user of "sensor calibration," "faulty sensor," "high alarm," or similar informational phrases. Manufacturers seem to be trying to close the generation gap that continues to exist between the world of science fiction and the real world.

The eternal triangle between on-line analyzers, process control instruments, and computers continues to grow and mature, and so it should.

GLOSSARY

ACTIVE ELEMENT (WHEATSTONE BRIDGE). The arm in which a resistance sensor is connected.

AIR ASPIRATOR. A device such as a Venturi that uses a high-velocity air stream to develop low pressure at the venturi's vena contracta and thus induce flow in another stream connected to the venturi at the low-pressure point. These devices are used frequently as the motive force for sample systems that handle air pollutants or stack emissions. See "Water Aspirator."

ALUMINUM OXIDE SENSOR. A hygrometer sensor consisting of porous gold and aluminum electrodes separated by an aluminum oxide dielectric.

ANALYZER HOUSE. A single- or multiple-room enclosure specially designed for analyzers and their support facilities. These buildings can provide efficient distribution of the many specific utility supplies required by analyzers, an environmentally protected atmosphere, and more centralized communication between groups of analyzers and the operating or maintenance personnel responsible for them.

BACKFLUSH. Generally refers to reversing the direction of carrier gas flow through a gas chromatograph column. The purpose of backflushing is to stop any further permeation of the column by the previously injected components and carry them out of the column either to a vent, another column, or the detector.

BACKPRESSURE. The fluid pressure developed by a flow-modulating device that maintains a constant inlet pressure. This helps to stabilize pressure variations in the samples to some analyzer detectors that must operate at a constant, higher pressure than the downstream sample discharge pressure.

BLIND FLANGE. A solid flange generally used to seal off a similarly sized flanged nozzle. Blind flanges may be drilled and tapped or welded to provide a mounting for analytical probes.

CCR. Abbreviation for central control room; the location where the majority of process control stations are grouped or where a host or supervisory computer is located.

CGA FITTINGS. Compressed-gas cylinder pressure regulator inlet fittings approved by the Compressed Gas Association for specific cylinder-to-regulator connections. The fittings prevent accidental mismatching between cylinder contents and the system downstream of the regulator. Fuels and oxidants, toxic and nontoxic, or corrosive and noncorrosive combinations are precluded by allowing only certain cylinder-to-regulator connections.

CALIBRATION GAS. Generally a 500- to 2000-psi pressurized gas cylinder filled with a mixture of gases containing one or more components of interest.

257

CARRIER GAS. The flowing gas that provides the transport mechanism for sample components through a chromatographic column.

CARTRIDGE FILTER. An assembly consisting of a body or holder with a replaceable internal filter cartridge.

CATALYTIC GAS DETECTOR. Refers to a catalyst-coated resistive sensor inside an ambient air and vapor detector cell. The catalytic sensor is heated to a temperature sufficient to oxidize all combustible gases in close contact with it. The coated sensor is usually an active element in a Wheatstone Bridge circuit. Ignition of combustible vapors outside the detector cell is prevented by a flame arrestor in the vapor entry pathway of the detector cell.

CELL CONSTANT. The ratio of electrode spacing in centimeters to the electrode area in square centimeters. The unit for the cell constant is thus 1/cm.

CELL WINDOW. A transparent or semitransparent material used with optical detector cells in photometric analyzer systems. Quartz, Pyrex®, sapphire, calcium fluoride, and barium fluoride are common cell window materials.

CHOPPER WHEEL. A motor-driven wheel containing two or more optical filters having transmittance bandpasses specifically chosen for a measuring or reference wavelength. Choppers often use synchronization elements such as imbedded magnets or machined slots to enable analyzer circuit discrimination of the transmitted energy beam of each filter.

CLOSED-LOOP CONTROL. Direct process control as opposed to process monitoring or open-loop indirect control. For example, analyzer output signals can be used as process variables for controllers that automatically regulate other mechanisms such as final control elements or valves. The effect of the final control element on the analyzer's sample and its output signal constitutes a form of feedback or closed-loop flow or process control information called closed loop control.

COALESCING FILTER. A porous filter media that causes small liquid droplets or mists in the sample gas stream to form larger liquid drops. The larger drops of liquid tend to seep (because of their weight) to the bottom of the filter medium and fall into a trap or similar mechanism that removes the liquids from the filter.

COLORIMETER. An analyzer that measures color purity or color intensity in accordance with standard color scales. Some chemical titrators operate with colorimeter detectors that are sensitive to sudden or specific color changes.

CONTROL VALVE TRIM. The plug-and-seat combination of a control valve; the size and shape of the trim determines the maximum flow and flow characteristics of a particular valve.

CORROSION ALLOWANCE. An additional amount of material used for equipment or pipe fabrication equal to an amount lost because of the expected corrosion rate for new equipment operating under normal process conditions.

CRACKING PRESSURE. The lowest pressure at which a relief valve will open and permit flow.

CRYSTAL OSCILLATOR HYGROMETER. A hygrometer sensor consisting of a hygroscopically coated crystal whose frequency and amplitude of vibration change as a function of absorbed moisture.

DIAMAGNETIC MATERIALS. Materials having a permeability slightly less than 1.0. These materials are slightly repelled by a magnetic field and are opposite in nature to paramagnetic materials.

DIATOMACEOUS EARTH. A finely divided white powder composed of the skeletons of micron-sized prehistoric sea animals called *diatoms*. This material is sometimes used as filter media, fine polishing compound, or chromatographic column support.

DIRECT-INSERTION PROBE. Any probe that can be inserted directly into a process line to measure or obtain information about a process variable. This term is generally used for analyzer sample collecting devices inserted into flowing process streams or into process vessels.

DOUBLE BLOCK VALVES. Two block valves installed in series with the intent to provide tight shutoff or leakproof stream selection of process material. Double block and vent valve pipe networks provide additional protection against possible leak-through by not allowing any internal pipe pressure to develop between the block valves.

DRIFT. See "Zero Shift."

DUAL FILTER. Two filters usually installed in parallel with suitable isolation valves. One filter is normally placed on line with the other filter in standby. The standby filter is valved in when the on-line filter needs to be cleaned.

EPROM. Abbreviation for erasable programmed read-only memory. This semiconductor memory is generally programmed by equipment manufacturers for specific applications that require slightly different control routines. Examples of this are the variations in titrator, pyrolyzer, or chromatographic analytical sequences now automated with digital memory systems.

EFFLUENT. A process stream leaving the plant or manufacturing area. Also, the vapor mix of carrier gas and sample components exiting from a chromatographic column.

ELECTRICAL RESISTANCE (ER) CORROSIMETER. A type of corrosion monitor that measures the average corrosion rate of an electrochemical system over a discrete period of time.

ELECTRICAL HEAT TRACING. Electric heat-tracing elements embedded in insulated fiberglass tape that can be used to heat process or sample stream piping. The heat tracing is wrapped around or routed along the side of the pipe or tube system. Electrical control can be by current limiting features of the tracing or by electrical temperature controller.

ELECTROCHEMICAL POLARIZATION. The reduction potential of an electrochemical reaction such as $Ag^+ + e^- = Ag$ at $+0.7996$ V DC.

ELECTRODE. One terminal in contact with an electrical source such as an anode or cathode.

260GLOSSARY

ELECTRODE FOULING. Any electrode coating or film that prevents normal electrode operation by either separating the electrode from fluids in close proximity with it or by causing changes in the resistance of the electrode–fluid contact junction.

ELECTROLYTE. Any chemical compound, acidic or basic, that forms ions when dissolved in water. Ions in solution conduct electrical current.

ELECTROLYTIC HYGROMETER. A hygrometer sensor consisting of two electrodes with a phosphorous pentoxide desiccant coating. The sensor electrolyzes water in the sample, and its amplifier measures this electrolysis current. This current is proportional to the amount of water that the desiccant has absorbed.

ELUTION. The removal of adsorbed or absorbed components from a gas chromatograph analytical column in order of their retention time. Components that are not strongly retained have the shortest retention time and are carried out of the column first.

EQUILIBRIUM CONDITION. In the case of hygrometers, this term refers to electrolysis current being directly proportional to the amount of moisture present in a gas stream (Chapter 5). Analytical equilibrium is usually the time at which chemical balance is achieved in titrations, chromatographic columns, and some ion-sensing electrodes. Mechanical equilibrium is a force balance neutral state where input forces are balanced by opposing feedback forces.

ERROR MESSAGES. A computer diagnostic communication language transmitted to printers and other similar display terminals to advise the user as to the type and location of central processor or peripheral device faults. Messages can also appear in abbreviated form such as code on DVMs or specially formatted CRT displays.

ESCALATION FACTOR (COST). A cost factor intended to compensate project estimates for the cost of living and cost of material increases likely to occur before project completion.

EXPLOSIONPROOF. The term used to describe equipment enclosures that are capable of withstanding an internal explosion without igniting a flammable gas surrounding the enclosure.

FILTER–SEPARATORS. A device used to separate entrained liquids and solids from gas in process or sample streams; also a coalescing filter.

FINAL CONTROL ELEMENT. The last element in a closed or feedback control loop. Usually a control valve but could be a pump stroke adjustment, vane control, or other similar mechanical device capable of moderating or varying the process variable as required to achieve equilibrium between controller setpoint and PV input. This "closes the loop" and is the essence of feedback control loops.

FLOW TRANSMITTER. A device that produces an output signal proportional to an input analog such as differential pressure acoss a flow orifice representing the measured flow.

FLOW-THROUGH CELL. An arrangement of pH, conductivity, or other similar ionic probe mounted in a fixed holder. The sample stream is piped to the probe holder and flows past the probe, returning to process, vent, or drain.

FULL-PORT BALL VALVE. A valve that permits unrestricted flow of process material when open and provides tight shutoff when rotated 90°. Direct insertion probes may be mounted through these types of valve.

FUME BARRIER. A seal between two rooms or different pieces of equipment in an analyzer house. The fume barrier isolates potentially harmful fumes from operating equipment and/or personnel.

GAS DIFFUSION COLUMN. A tubular device designed to promote blending or mixing of two or more gas streams. Certain types of these columns are used as calibration standards for sulfur analyzers.

GAS SCRUBBER. A liquid-filled chamber through which sample gas can be bubbled to remove soluble contaminants or particular matter. The chamber may be packed with glass beads, ceramic rings, or Teflon® shavings to increase the gas–liquid contact surface area.

GEL-FILLED REFERENCE ELECTRODE. A long-life type of pH reference electrode designed to be a replacement for older-type pH reference electrodes that were filled with a saturated potassium chloride solution. The gel types have a permanent electrolyte fill.

GROUND. Earth ground reference for equipment, structures, and instrument circuits. This is usually accomplished with a direct electrical connection to a metal rod driven down to and in contact with the highly conductive groundwater table. An underground interconnection grid is often included in the site construction of petrochemical plants to provide all process equipment, instrumentation, and supporting structures with a common, single-potential ground reference.

GUARD CIRCUIT. A type of voltage feedback circuit used to electrically drive a coaxial shield at the same potential as the center conductor. This minimizes electrochemical polarization of ion-sensing electrodes and permits the use of longer coaxial cable lengths.

HEAT TRACING. Any steam tubing or electrical heating wire system used for heating process or sample lines to prevent freezing or condensation of liquids in vapor phases. Heat tracers are usually wrapped around or placed under the line to be heated and covered with insulation.

HEAT-TRANSFER CEMENT. A heat-conducting bonding putty used to transfer heat from heat tracing to process or sample lines; also known as *heat-transfer material* (HTM).

HIGH IMPEDANCE. The characteristic of high resistance to input current signals at the input terminals of a preamplifier or amplifier circuit. An amplifier with high input impedance permits accurate monitoring of probe or sensor signals because of the low current loading effect this impedance imposes on the input device.

HIGH-SPEED BYPASS LINE. A high-flow-velocity sample line used to bypass a portion of sample flow prior to entry into an analyzer. Inlet sample flow requirements of most on-line analyzers are so small that additional bypass flow must be used to reduce the sample residence time to the analyzer.

HYGROSCOPIC COATING. The surface of objects coated with a porous, water-absorbing material.

I/P CONVERTER. An electropneumatic instrument that converts a DC current input signal (e.g., 4 to 20 mA DC) to a pneumatic output signal (e.g., 3 to 15 psig). These transducers are typically used when an electronic panel instrument must output a signal to an air-operated control valve.

INDIRECT CONTROL. Refers to a process control configuration using analyzers or other input devices to provide a setpoint or advisory function to a master control device such as a process control computer.

IN-LINE ANALYZER. Refers to an analyzer whose sensor is mounted directly in the process stream for continuous, real-time analysis of a specific process stream component or characteristic such as pH or conductivity.

INSERTION PROBE. An ion-sensing, conductance, or sample-gathering probe that can be installed directly in a process line and sealed with a tubing fitting at a threaded or flanged connection.

INTERFERENCES. Chemical compounds that have characteristics similar to the components of interest monitored by a particular analyzer. These similarities can cause analyzer detectors to respond equally to measured components and interfering compounds, thereby invalidating the analysis.

INTERLOCK. A device that interrupts an instrument system, machine, or process when abnormal operation is sensed in order to prevent unsafe operation and thus protect equipment or personnel.

INTRINSIC SAFETY. Refers to limiting the electrical discharge potential of a device so that neither a spark nor sufficient heat can be generated by the device and make it unsafe in a hazardous environment.

LABORATORY ANALYZER. A versatile, bench-mounted analytical instrument designed for the nonhostile laboratory environment.

LAMINAR FLOW. A type of nonuniform flow in a pipe that can cause stream components to be separated as a result of differences in viscosity, density, or other physical variables. Laminar flow is usually undesirable because of measuring errors introduced across flow elements designed for turbulent flow.

LIGHT-SCATTERING PHOTOMETER. Another name for a turbidimeter. Surface, right-angle, or forward light-scattering techniques may be employed. Forward scatter photometers are sometimes called *suspended-solids monitors* (SSM).

LIMITING ORIFICE. Usually an orifice installed between piping flanges to limit the maximum flow of process material. See "Restricting Orifice."

LINEARITY. The measure of the output signal compatibility of an instrument as compared to an output calibration "curve" of specific slope and mathematically representing a straight line.

MIR. Abbreviation for multiple internal reflection. Refers to an IR absorption technique used with dirty liquids or highly concentrated absorbing compounds.

MV/I. Abbreviations for the terms "millivolt" and "current." Also refers to an electronic device that converts a low-level DC voltage (e.g., 0 to 10 mV DC) to a low-impedance DC current signal (e.g., 4 to 20 mA DC). These converters are often used to transmit analyzer or thermocouple signals from field locations to a central control room. See "Transducer."

MAGNETIC WIND. A descriptive phrase representing the rapid flow or diversion of an oxygen-bearing gas stream within a ring-type, thermomagnetic oxygen sensor.

MASS FLOW. The product of density times a constant and volumetric flow. Units are usually in pounds per hour, abbreviated as pph.

METER PROVER. A generalized name for signal simulators and volumetric flow meter test equipment. pH, moisture, corrosion, or other sensor outputs can be generated by signal simulators, and in the case of volumetric meter provers, a plug or "pig" is cycled through a known volume of pipe networks to account for an exact displacement of the internal fluid.

METERING VALVE. A precision flow control valve usually equipped with a long tapered plug and small-diameter seat ring or orifice. See also "Needle Valve."

MICROPROCESSOR. A family of LSI semiconductor devices consisting of logic elements such as central processor, arithmetic logic unit, and various input–output registers. They accept digital input data, shifting or otherwise logically manipulating it through programming sequences, and producing digital outputs. Also, any stand-alone or on-board single-chip computing system used to increase the versatility and flexibility of electric typewriters, duplicating machines, on-line analyzers, controllers, and many other discrete instruments.

MIXING TEE. A piping tee where two streams or flow components blend together and combine with sufficient velocity and turbulence to promote thorough mixing in a common outlet stream.

NEEDLE VALVE. A manual or automatic valve featuring a long, tapered plug capable of setting accurately metered flow. See also "Metering Valve."

NONPOLAR LIQUID. A liquid that does not readily ionize or contain a significant number of ions. In chromatography, the term "nonpolar" is used to represent molecules that have symmetrical charge distribution or low hydrogen bonding activity.

OCTANE ANALYZER. An analyzer designed to determine the research octane number (RON) of engine-rated gasoline standards.

OFFSET CURRENT. The output current of a differential amplifier that is present when both inputs are short-circuited or connected to signal ground.

ON-LINE. Usually refers to continuous, real-time service or operation as opposed to intermittent operations. Electronic data processing peripherals are said to be "on line" when they are operating in their normal communication mode with a central processor.

PMT. Abbreviation for photomultiplier tube. It amplifies low-level light signals and produces a proportional output current.

PROM. Abbreviation for programmed read-only memory. Refers to semiconductor memory elements that contain programming that cannot be altered by the user.

PARAMAGNETIC MATERIALS. Materials having a permeability slightly greater than 1.0. These materials are slightly attracted by a magnetic field. The opposite of diamagnetic materials. See "Diamagnetic Materials."

PARTITION COEFFICIENT. A ratio representing the distribution of solute molecules between phases in a chromatographic column.

pH CHARACTERIZATION. Refers to application of a nonlinear, variable-gain controller to provide more or less uniform sensitivity to pH changes over the pH control range.

PHOTOMETRIC ANALYZER. Generally a light-measuring analytical system operating in the UV, IR, or visible band of the electromagnetic spectrum.

PNEUMATIC CONTROLLER. A controller powered by an air supply that produces a pneumatic or air output signal proportional to the difference between the controller's setpoint and the process variable input.

PRECONDITIONING. The initial conditioning of a sample before it enters the primary sample system.

PRESSURE LETDOWN VALVE. A pressure control valve generally used to reduce higher upstream process pressure to levels acceptable for constant downstream pressure applications.

PREVENTIVE MAINTENANCE. Certain maintenance procedures performed on many types of equipment such as analyzers, instrumentation, and mechanical control devices to prolong their service life and increase minimum time between failures. With preventive maintenance, work is done on a prearranged schedule as compared to a routine maintenance program where work is done periodically as needed.

PROBE SIMULATOR. Usually an electrical device that can produce an output signal identical to the signal produced by a pH, conductivity, corrosion, or other similar sensor.

PROCESS HEADER. A major process pipe with numerous smaller lines removing or returning process material to the header.

PROCESS STREAM. Any fluid stream in the plant or manufacturing process that contains process influent or effluent.

PROCESS STREAM COMPONENTS. Usually chemical variables in a process stream. Sometimes used to represent different phases or physical characteristics of a flowing stream.

PROCESS TAP. The location of a sample line or smaller process line opening into a process vessel or header. "Process tap" and "sample tap" are terms sometimes used interchangeably.

PROCESS VARIABLE. Any dynamic characteristic of an operating unit such as flow, pressure, temperature, or chemical concentration that can change under certain process conditions. The control of these variables tends to control the process.

PROMPTING ROUTINE. A computer communication format that advises the user of the next entry step to be initiated for further data processing.

PURGE GAS. Any inert gas used to remove residual process gas or liquid from a pipe or tubing system. Also, a safe, inert gas used for continuous replacement of instrument enclosure atmospheres.

RAM. Abbreviation for random access memory, also called *read/write memory*. This type of memory is generally used for applications that require changes in range—the total input or output differential capability of an instrument. See "Span."

RFI. Abbreviation for radio frequency interference. A type of electrical interference that can be produced by mobile radio transmitters and picked up by poorly shielded electronic circuitry.

RECOMBINATION EFFECTS. Refers to the reformation of water by the combination of oxygen and hydrogen after initial separation by electrolysis.

REFLECTED ENERGY TRANSMISSION. Refers to radiated IR energy passing back and forth between highly polished cell interior walls of IR analyzer systems.

REFRACTOMETER. An analyzer that determines the critical angle of incidence at which light is no longer refracted by a liquid, but is reflected.

RELIEF VALVE. An overpressure protection device that diverts process or sample line contents to lower-pressure systems.

REMOTE MULTIPLEXING. An instrumentation system design for controlling or monitoring multiple sensors with several distant sequencing networks that communicate with a single control computer or microprocessor.

REPRESENTATIVE SAMPLE. A sample physically and chemically identical to the process material it was taken from.

RESIDENCE TIME. The time delay attributed to the transport of sample material from the sample tap to the detector of the on-line analyzer.

RESPONSE TIME (OF INSTRUMENT). The time required for an output signal of an instrument to reach 63% of its final value when its input is subjected to a step change in value.

RESPONSE TIME (OF PROCESS). The time delay attributed to the transport of the process material from the point of origin of the process change until it reaches the sample tap of the analyzer.

RESTRICTING ORIFICE. An orifice used to reduce or inhibit sample system or process flow. Often a large pressure drop may be taken across a restricting or limiting orifice to reduce downstream pressures.

RETENTION TIME. The time from sample injection to component elution in chromatography.

RETRACTABLE PROBE. A probe that can be removed from a process while the process is operating at normal pressure.

REVERSE-FLOW CHECK VALVE. Usually a valve with a spring-loaded ball or poppet that prevents reverse flow of process or sample material between its inlet and outlet ports.

ROTAMETER. A vertical flow metering device usually consisting of a tapered glass tube enclosing a "float" or plummet. The plummet develops a constant pressure drop versus flow rate, thus seeking a tube internal area, allowing upward stream flow forces to balance plummet weight.

ROUTINE MAINTENANCE. Maintenance performed on an as-required basis. Routine inspections may indicate the need for routine maintenance work as compared to preventive maintenance where certain maintenance work may be done even though the equipment appears to be in satisfactory operating condition.

RUPTURE DISK. A thin, preformed metal diaphragm installed immediately underneath a relief valve. It is generally used to prevent relief valve internals from coming in contact with corrosive process fluids.

SAMPLE BLOCK VALVE. Any valve used to isolate a sample line from a process header.

SAMPLE BOOTH. A small enclosure designed to safeguard personnel while hazardous process samples are collected. These enclosures usually have splash shields, fume hoods, chemical drains, and emergency first-aid equipment located nearby.

SAMPLE CELL. The measuring chamber of an analyzer. In photometric systems, the sample cell is a compartment with windows through which a controlled light path passes through the sample. In potentiometric or amperometric systems, the sample cell is an electrode holder with sample flow passages.

SAMPLE CYLINDER. Usually a small metal cylinder used to collect a representative sample of process material for safe transport to remotely located analysis instrumentation.

SAMPLE PRECONDITIONER. The tubing, filters, valves, and pressure regulating devices necessary to produce a clean, representative sample for transport to the on-line analyzer system.

SAMPLE SYSTEM. All tubing, filters, valves, and other regulating or safety devices required to provide a continuous or intermittent flow of sample material to an on-line analyzer and from there to some final destination.

SAMPLE SYSTEM TAP. The point where a representative sample is withdrawn from a process header for subsequent transport to the sample system.

SCALE FACTOR. Refers to converting a standard output current or voltage to engineering units. May also be associated with the gain or calibration circuits of an instrument.

SELF-DIAGNOSTICS. A software program built into some computers and microprocessors to increase the ease of troubleshooting and minimize instrument downtime. Provides an indication as to the location of a fault or what type of maintenance procedures should be executed.

SENSOR. Any device capable of detecting one or more process variables such as flow, pressure, temperature, pH, conductivity, or component concentration.

SPAN. A particular input or output differential within the range of an instrument. Also, the slope of the calibration curve of the instrument. The terms "span," "gain," and "sensitivity" are sometimes used interchangeably. See "Range."

SPECIFIC CONDUCTANCE. The conductivity of a 1-cm cube of solution at 25°C and atmospheric pressure.

SPECIFIC RESISTANCE. The resistance of a 1-cm cube of solution at 25°C and atmospheric pressure.

STANDARDIZE CONTROL. A control commonly used with pH monitors to compensate for probe aging. The output or display meter is adjusted with the standardize control to read the same as a known pH buffer solution in which the electrodes are immersed.

STATIONARY PHASE. Refers to the liquid-coated column packing material in gas chromatography.

STEAM TRACING. A steam tubing system used to heat a process or sample line. See "Heat Tracing."

STREAMING POTENTIAL. A nonuniform potential sensed by pH electrodes in some nonuniform, highly conductive streams found in plating baths and in fast-flow, high-purity water streams from deionization systems.

SUBMERSION PROBE. A pH, conductivity, or other probe usually attached to a long pipe handle and totally immersed in an open process vessel or natural stream.

TITRANT. A reagent chemical of known strength used to determine some chemical characteristic of an exact quantity of solution under test. The titrant is slowly added to the solution being tested until a desired endpoint, color change, or pH level if reached.

THERMOMAGNETIC OXYGEN SENSOR. Usually a thermal conductivity bridge within a superimposed magnetic field. The TC sensor responds to the changes induced in the sample by the magnetic field.

THRESHOLD LIMIT VALUE (TLV). The concentration of hazardous or toxic substances in air that can be safely tolerated on a daily basis.

THROTTLING VALVE. A manual or automatic control valve designed to regulate the pressure drop or the flow or process fluid through the valve.

TRACK-AND-HOLD FUNCTION. A capacity memory circuit used to store the most recent voltage signal from a particular circuit until some internal analyzer condition (such as automatic calibration or photometer beam switching) is concluded.

TRANSDUCER. A pneumatic, electromechanical, or electronic instrument used to convert flow, pressure, temperature, or analytical engineering units into another signal level or signal type. See "mV/I."

UV–VIS. Abbreviation for ultraviolet–visible; a particular band of the electromagnetic energy spectrum used in some photometric analyzers.

UNATTENDED OPERATION. A desirable but difficult-to-achieve type of operation when used in conjunction with on-line analyzer systems. The term generally implies that the instrument has excellent long-term drift and stability characteristics.

UNIVERSAL OUTPUT. Any electronic output circuit characterized by both current and voltage outputs in field selectable ranges such as 4 to 20 mA DC or 0 to 10 V DC.

VENTED SAMPLE SYSTEM. Any sample system that exhausts the sample stream to a vent header instead of returning it to the process.

VENTURI. A flow element consisting of oppositely tapered cones with the small ends connected together. A flow constriction called a *vena contracta* is produced at the high-velocity point through the venturi with a resultant drop of pressure at the constriction.

VOLUMETRIC CHAMBER. A controlled volume chamber placed in a process pipe or sample tubing line to promote the mixing of two or more stream components.

WATER ASPIRATOR. A device using a high-velocity water stream flowing through a venturi to create a low pressure at the vena contracta and thus induce flow in a low-velocity stream connected at that point. See "Air Aspirator."

ZERO. Refers to the lower limit of a particular input or output or span; for instance, the zero of a 4- to 20-mA DC output signal is 4 mA DC.

ZERO SHIFT (OR DRIFT). An error characterized by a parallel shift of the calibration curve of an instrument. This causes the lower limit of the calibration curve to originate at an incorrect zero point.

ZIRCONIUM OXIDE CELL. A type of oxygen sensor consisting of platinum electrodes on opposite sides of a zirconium oxide dielectric crystal. The crystal lattice becomes a semiconductor for oxygen ions when it is heated to relatively high temperatures. A millivoltage is developed across the crystal proportional to the difference of oxygen partial pressures applied to each side of the lattice.

APPENDIX

1

LIST OF TABLES AND FIGURES

269

APPENDIX

2

ALPHABETICAL LISTING
OF MANUFACTURERS

Mfgr. No.	Manufacturer	Product(s)
1	Ametek Thermox Instruments Division 150 Feeport Road Pittsburgh, PA 15238 Telephone: (412) 828-9040	Oxygen and combustible stack gas analyzers
2	AMSCOR 212 N. Velasco Angleton, TX 77515 Telephone (713) 849-8462	Chromatographs: slider-type valves
3	Analytical Instruments Corp. 8250 West Little York Road Houston, TX 77040 Telephone: (713) 466-6105	Chromatographs: 10-port valve
4	Anarad, Inc. P. O. Box 3160 Santa Barbara, CA 93105 Telephone: (805) 963-6583	Infrared analyzers for gas and liquid service
5	Applied Automation, Inc. A subsidiary of Phillips Petroleum Pawhuska Road Bartlesville, OK 74004 Telephone: (918) 661-6141	Chromatographs: plunger-type valves
6	Bacharach Instrument Co. 625 Alpha Drive Pittsburgh, PA 15238 Telephone: (412) 782-3500	Gas-detection systems

Mfgr. No.	Manufacturer	Product(s)
7	Badger Meter, Inc. Precision Products Division 6116 East 15th Street Tulsa, OK 74112 Telephone: (918) 836-8411	Research control valves: sample system components
8	Balston, Inc. P. O. Box C 703 Massachusetts Ave. Lexington, MA 74112 Telephone: (617) 861-7240	Filters, sample system components
9	Beckman Instruments, Inc. Cedar Grove Operations 89 Commerce Road Cedar Grove, NJ 07009 Telephone: (201) 239-6200	Chromatographs; conductivity, oxygen, and dissolved oxygen analyzers
10	Collins Products Company P. O. Box 282 Livingston, TX 77351 Telephone: (713) 327-4200	Sample systems and sample system components
11	DuPont Analytical Instruments Concord Plaza Wilmington, DE 19898 Telephone: (302) 772-5481	Liquid chromatographs, leak detectors, UV emission, and stack gas analyzers; moisture analyzers
12	The Electron Machine Corp. P. O. Box M Umatilla, FA 32784 Telephone: (904) 669-3101	Critical-angle refractometers
13	Filterite Corporation A Brunswick Subsidiary Timonium, MD 21093 Telephone: (301) 252-0800	Cartridge-type filters and sample system components
14	Fischer & Porter Company 1000 Warminster Road Warminster, PA 18974 Telephone: (215) 674-6000	Chlorine detectors, aperometric analyzers, dissolved oxygen, pH, and redox meters

Mfgr. No.	Manufacturer	Product(s)
15	Fisher Controls Company Marshalltown, Iowa 50158 Telephone: (515) 754-3011	Sample system components
16	Fluid Data, Inc. 1844 Lansdown Ave. Merrick, NY 11566 Telephone: (516) 223-2190	Colorimeters and pyrolysis gas sample conditioners
17	Foxboro Analytical 140 Water Street South Norwalk, CT 06856 Telephone: (203) 853-1616	Chromatographs; IR, octane, ferrograph, particle size, pH, conductivity, and ORP analyzers
18	Go, Inc. 11940 E. Washington Blvd. Whittier, CA 90606 Telephone: (213) 945-3691	Pressure regulators, back pressure regulators, and sample system components
19	Hack Company P. O. Box 389 Loveland, CO 80537 Telephone: (303) 669-3050	Pump colorimeters and turbidimeters
20	IBM Instruments, Inc. P. O. Box 332 Danbury, CT 06810 Telephone: (203) 796-2500	Polarographic–voltametric analyzers
21	Ionics, Inc. 65 Grove Street Watertown, MA 02172 Telephone: (617) 926-2500	Automatic titrators; total carbon and total organic carbon analyzers
22	ITT Barton P. O. Box 1882 City of Industry, CA 91749 Telephone: (213) 961-2547	Densitometer systems
23	Leeds & Northrup Co. Sunnytown Pike North Wales, PA 19454 Telephone: (215) 643-2000	Oxygen, TC, IR, pH, conductivity, redox, specific ion and suspended solids analyzers

Mfgr. No.	Manufacturer	Product(s)
24	Milton Roy Company Hays-Republic Division 4333 S. Ohio Street Michigan City, IN 46360 Telephone: (219) 879-4441	Colorimeters; oxygen, pH, conductivity, combustibles, thermal conductivity and dissolved oxygen analyzers
25	Mine Safety Appliance Co. 600 Penn Center Blvd. Pittsburgh, PA 15235 Telephone: (412) 273-5000	IR, oxygen, combustibles, thermal conductivity and total hydrocarbon analyzers
26	Moisture Systems Corp. 120 South Street Hopkinton, MA 01748 Telephone: (617) 435-6881	IR-type moisture in solids analyzers
27	National Sonics Division Xertex Corporation 250 Marcus Blvd. Hauppauge, NY 11787 Telephone: (516) 273-6600	Polarographic instruments for dissolved oxygen, residual chlorine, and chlorine dioxide
28	Nupro Company 4800 East 345th Street Willoughby, OH 44094 Telephone: (216) 951-7100	Sample system components
29	The Ohmart Corp. 4241 Allendorf Drive Cincinnati, OH 45209 Telephone: (513) 272-0131	Radiation moisture and density analyzers
30	Ondyne, Inc. P. O. Box 6302 Concord, CA 94524 Telephone: (415) 825-8282	Aluminum oxide type hygrometers
31	Orbisphere Laboratories 20902 S. Brookhurst Suite 221 Huntington Beach, CA 92646 Telephone: (714) 964-2410	Dissolved oxygen analyzers

Mfgr. No.	Manufacturer	Product(s)
32	Panametrics, Inc. 221 Crescent Street Waltham, MA 02154 Telephone: (617) 899-2719	Moisture analyzers
33	Petrolite Instruments 411 Bluebonnet Drive, Suite C Stafford, TX 77477 Telephone: (713) 494-7121	LPR corrosion monitors and conductivity instruments
34	Rohrback Instruments 1181 E. Telegraph Rd. Santa Fe Springs, CA 90670 Telephone: (213) 949-0123	ER and LPR corrosion monitors
35	Scientific Gas Products 2330 Hamilton Blvd. P. O. Box 312 S. Plainfield, NJ 07080	Calibration gases and gas safety data
36	Siemens, Inc. M & C Section E681 Postfach 21 1080 D-5000 Karlsruhe 21 West Germany	IR, moisture, oxygen, pH, conductivity, redox, dissolved oxygen, and FID analyzers
37	Signet Scientific 3401 Aerojet Avenue El Monte, CA 91734 Telephone: (213) 571-2770	pH, conductivity and ORP analyzers
38	Solartron Transducer Group Jantech Corporation 114 State Street Boston, MA 02109	Specific gravity gas and liquid density analyzers
39	Taylor Instrument Co. 95 Ames Street Rochester, NY 14601 Telephone: (716) 235-5000	Paramagnetic oxygen analyzers

Mfgr. No.	Manufacturer	Product(s)
40	Texas Analytical Controls P. O. Box 42520 Houston, TX 77042 Telephone: (713) 491-4160	Hydrogen sulfide and combustible gas detectors and monitors
41	Thornton Associates, Inc. 87 Beaver Street Waltham, MA 02154 Telephone: (617) 899-1400	Conductivity and resistivity monitors
42	Uniloc Division Rosemount, Inc. 2400 Barranca Rd. Irvine, CA 92714 Telephone: (714) 546-8700	pH, conductivity, redox, and residual chlorine analyzers
43	Van London Company, Inc. 6103 Glenmont Houston, TX 77081 Telephone: (713) 772-6641	pH and conductivity sensors, analyzers and custom packaging
44	Westinghouse Electric Corp. Combustion Control Division 1201 North Main Street Orrville, OH 44667 Telephone: (216) 682-9010	Stack gas oxygen and combustibles analyzers
45	Yellow Springs Instrument Co. P. O. Box 279 Yellow Springs, OH 45387 Telephone: (513) 767-7241	Salinity, temperature, and polarographic O_2 analyzers

APPENDIX

3

REFERENCES

CHAPTER 2

1. R. Strauss, Sample Filters for Analyzers, *Instrumentation Technology,* August 1977.
2. Bulletin 17-4A, Sample Dilution System for High Concentration Analysis, Fischer and Porter, 1977.
3. Bulletin 70-9001, Technical Information for Handling Chlorine, Sulfur Dioxide and Ammonia from Supply to Point of Application, Fischer and Porter, 19
4. Bulletin NW-273, Check and Relief Valves, Nupro Company, Markad Service Company, 1981.
5. *Control Valve Engineering Handbook,* Foxboro, 1981.

CHAPTER 3

1. M. S. Frant and R. T. Oliver, *Process Analytical Measurements,* American Chemical Society, 1980.

CHAPTER 4

1. Bulletin 17-7, pH Hydrogen Ion Concentration and Buffering, Fischer and Porter, 1979.
2. *Introduction to pH,* Beckman Technical Information, 1977.
3. D. L. Hoyle, The Effect of Process Design on pH and pIon Control, 18th ISA-AID Symposium, 1972.
4. D. C. Merriman, pH Probe Characteristics; an unpublished letter, 1981.
5. Marketing letter, Flow Powered Cleaner, Uniloc,® 1981.
6. C. C. Westcott and V. Kohler, Standard Techniques for pH Measurements with Troublesome Samples, Beckman Technical Information, 1978.
7. G. Shinsky, *The Dynamic Response of Potentiometric Electrodes,* Instruments and Control Systems, 1975.

8. J. V. Wright, Inductive Method of Electrodeless Conductivity Measurement, 1979 Symposium on Measurement Technology of the 80's.
9. R. F. Legenhausen, Electrolytic Conductivity Applications in Extractive Mettalurgy, 154th Meeting of the American Electrochemical Society, 1978.

CHAPTER 5

1. Bulletin H-2000, *The Newest Moisture Minder,* Panametrics, Inc.
2. M. Stern and A. L. Geary, Electrochemical Polarization, *Journal of the Electrochemical Society* (1957).
3. M. Stern and E. D. Weisert, Experimental Observations on the Relation Between Polarization Resistance and Corrosion Rate, *ASTM Proceedings* (1959).
4. R. R. Annand and P. E. Eaton, Modern Developments in Polarization Techniques, *Corrosion/73* (1973).
5. D. A. Jones and N. D. Greene, Electrochemical Detection of Localized Corrosion, 1969 NACE Conference.
6. C. R. Townsend, Comparison of 2 and 3 Electrode Polarization Resistance Methods of Corrosion Measurement, *Corrosion/74* (1974).
7. Booklet, Rohrback Instrument's Corrosion Monitoring Primer, Rohrback Instruments.
8. C. G. Arnold, Linear Polarization Instruments Mated with Process Control Computers, *Corrosion/76* (1976).
9. G. L. Cooper, Modern Microprocessor Electrical Resistance Instruments, International Corrosion Forum, 1982.
10. D. R. Berstrom, Chemical Process Control with Corrosion Sensors, *Corrosion/81* (1981).

CHAPTER 6

1. Bulletin B10007-880, Oxyprobe Combustion Efficiency, Milton Roy Co.
2. Bulletin 07-1013 DU, Magnetic Oxygen Analyzer Model 802, MSA Instruments.
3. Bulletin EG 3340-001-121, Oxymat 3 Oxygen Analyzer, Siemens Corporation.
4. P. W. Presepe, H. V. Mangin, and W. K. Barnett, CO/O_2 Combustion Control Strategy Provides Fuel Dollar Savings, ISA Capital Cities Control Conference, 1980.
5. Application Data Number 4386, Dissolved Oxygen Analyzer Applications in Water Pollution Control and Waste Treatment Plants, Beckman Instruments.

6. Instruction Manual for Model A-62606, Dissolved Oxygen Analyzer, Milton Roy Co., 1978.

7. R. Poole and J. Morrow, Improved Galvanic Dissolved Oxygen Sensor for Activated Sludge, *J. Water Pollution Control* (1977).

8. Bulletin PDS-A100B, Dissolved Oxygen and Temperature Analysis System, Delta Scientific Products, 1981.

9. Bulletin ADS 200, Continuous Dissolved Oxygen Control with Readout Module Mounted on Aeration Tank Rail, Delta Scientific Products, 1978.

10. Bulletin ADS-A201, Continuous Oxygen Control with Readout Module Mounted at a Location Remote to the Aeration Tank, Delta Scientific Products, 1978.

11. R. N. Salzman & P. J. Clark, Energy Efficient Aeration System, Mixing Equipment Co. & Leeds & Northrup Co., 1982.

12. J. M. Hale, Analysis of Trace Amounts of Dissolved Oxygen with the Orbisphere High Sensitivity Oxygen Measurement System, Orbisphere Laboratories.

CHAPTER 7

1. Bulletin 07-0518, Lira Model 303, Luft-type Infrared Analyzer, MSA Instruments.

2. Bulletin APS-R2, Infrared Analysis of Gas Process Streams and Liquid Process Streams, Foxboro, 1980.

3. M. S. Frant and G. LaButti, Process Infrared Measurements, *Analyt. Chem.* (1980).

4. Bulletin E681/1029-101, The Reliable Alternative: Ultramat 3, Siemens Corporation.

5. M. D. Rider, Monitoring Organic Pollutants Continuously by Total Carbon Analyzer (TCA), *Industrial Water Engineering* (1976).

6. Bulletin 07-0512B, Drying Oven and Solvent Recovery Monitoring Systems, MSA Instruments.

7. Bulletin 3A, Continuous Monitoring of Product Color with the Du Pont 400 Photometric Analyzer, Du Pont Instruments.

8. B. Reed, Theory, Calibration and Applications for the Model SSR-72 Electron Machine Corporation Critical Angle Refractometer: an unpublished letter, 1982.

CHAPTER 8

1. Instruction Manual for Digichem 3000 and 4000 Series Programmable Chemical Analyzers, Ionics, Inc., November 1980.

2. Barton's Densitometer: Probe Provides Outstanding Signal-to-Noise Ratio, *Instrumentation Insight* **8,** (2) (1981).

3. Bulletin, Liquid Density Transducers, Solartron Transducer Group.
4. Bulletin AD-100, Interface Detection, ITT Barton, 1981.
5. Bulletin AD-101, Jet or Diesel Engine Performance Testing, ITT Barton, 1981.
6. Bulletin PSS 6-9A1A, 81P Series Process Octane Analyzer, Foxboro, 1981.

CHAPTER 9

1. W. Ramsey, *Proc. Roy. Soc.* **A76,** 111 (1905).
2. M. Tswett, *Ber. deut. Botan. Ges.* **24,** 316, 384 (1906).
3. A. T. James and A. J. P. Martin, *Biochem. J. Proc.* **48,** vii (1951).
4. A. T. James and A. J. P. Martin, *Analyst* **77,** 915 (1952).
5. Compressed Gas Association, *User's Manual.*
6. J. J. vanDeemter, F. J. Zuiderweg, and A. Klinkenberg. *Chem. Eng. Sci.* **5,** 271 (1956).
7. D. E. Schup, *Gas Chromatography,* Wiley, New York (1968).
8. W. R. Supina, The Packed Column in Gas Chromatography, Supelco, Inc., Belefonte, Pa. (1974).
9. H. M. McNair and E. J. Bonelli, Basic Gas Chromatography, Varian Aerograph, Palo Alto, CA (1969).
10. W. A. Dietz, *J. Gas Chromatogr.* **5,** 68 (1967).
11. E. Kovats, *Helv. Chem. Acta* **41,** 1951 (1958).
12. A. Wehrli and E. Kovats, *Helv. Chem. Acta* **42,** 2709 (1959).
13. K. P. Hupe, *J. Gas Chromatogr.* **3,** 12 (1965).
14. C. L. Guillemin, Instrumentation Technology, 47 (August 82).
15. K. Unger et al., *J. Chromatogr.* 125 (1976).
16. J. J. Kirkland, W. W. Yau, and H. J. Stoklosa, *J. Chromatogr. Sci.* **15,** 303 (1977).
17. J. Q. Walker, M. T. Jackson, Jr., and J. B. Maynard, *Chromatographic Systems, Maintenance and Troubleshooting,* 2nd ed., Academic Press (1977).
18. L. R. Snyder and J. J. Kirkland, *Introduction to Modern Liquid Chromatography,* 2nd ed., Wiley-Interscience, New York (1979).
19. R. W. Yost, L. L. Ettre, and R. D. Conlan, *Practical Liquid Chromatography—An Introduction,* Perkin Elmer (1980).
20. J. B. Angell, S. C. Terry, and P. W. Barth, *Sci. Am.,* 50–53 (April 1983).

APPENDIX

4

RELATED READING MATERIAL

American Petroleum Institute, *Manual on Installation of Refinery Instruments and Control Systems,* 3rd ed., American Petroleum Institute, 1977

Callahan, F. J., *Swagelok® Tube Fitting and Installation Manual,* Crawford Fitting Company, 1974

Chemtrol® Plastic Piping Handbook, Celanese Piping Systems, Inc., 1973

Cornish, D. C., et al, *Sampling Systems for Process Analysers,* Butterworths, London, 1981

Fisher Control Valve Handbook, 2nd ed., Fisher Controls Company, 1977

Fontana, M. G., et al, *Corrosion Engineering,* 2nd ed., McGraw-Hill, New York, 1978

Liptak, B. G., *Instrument Engineers' Handbook,* Vol. 1, Chilton, 1969

APPENDIX

5

TRADEMARKS

Alloy 20 is a trademark of Carpenter Technology Corp.

Alumel is a trademark of Hoskins Manufacturing Co.

Carpenter is a trademark of Carpenter Technology Corp.

Chromel is a trademark of Hoskins Manufacturing Co.

Duranickel is a trademark of Huntington Alloys, Inc.

Grafoil is a trademark of Union Carbide Corp.

Hastelloy is a trademark of Stellite Division of Cabot Corp.

Inconel is a trademark of Huntington Alloys, Inc.

Kel-F is a trademark of 3-M Co.

Kynar is a trademark of Pennwalt Chemical Corp.

Monel is a trademark of Huntington Alloys, Inc.

Neoprene is a trademark of E. I. Du Pont de Nemours and Company.

Ni-Span is a trademark of Huntington Alloys, Inc.

Noryl is a trademark of General Electric Co.

Pyrex is a trademark of Corning Glass Works

Ryton is a trademark of Phillips Petroleum Co.

Teflon is a trademark of E. I. Du Pont de Nemours and Co.

Tefzel is a trademark of E. I. De Pont de Nemours and Co.

Tygon is a trademark of Norton Company.

Viton is a trademark of E. I. Du Pont de Nemours and Co.

FLOW CHARTS FOR GASES, LIQUIDS, AND STEAM

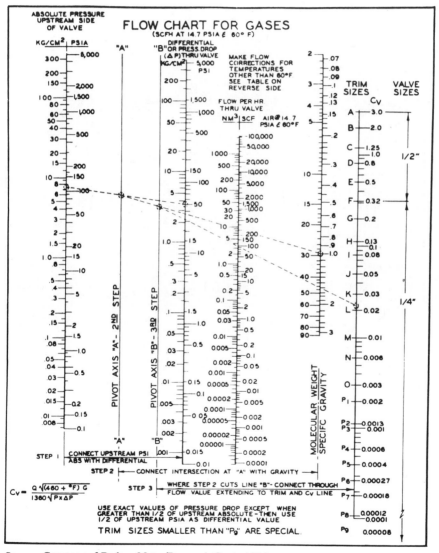

Source: Courtesy of Badger Meter/Research Control Valves.

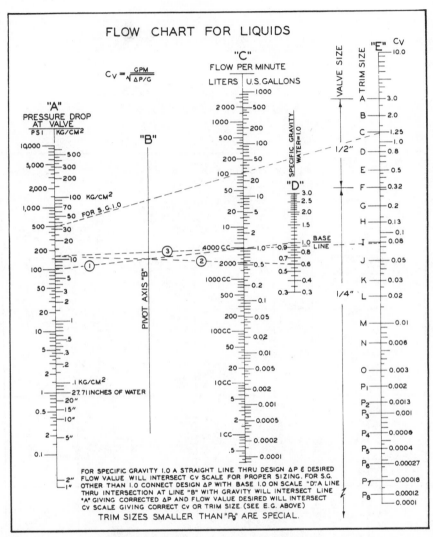

Source: Courtesy of Badger Meter/Research Control Valves.

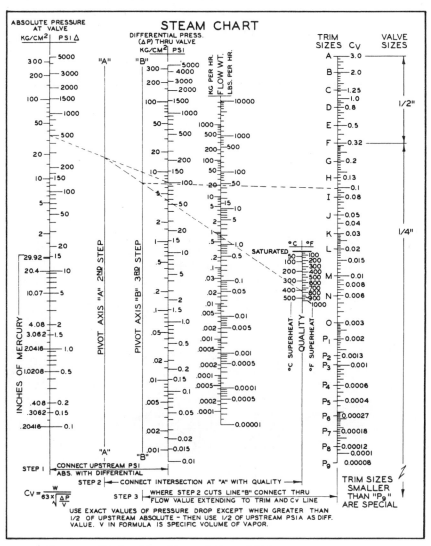

Source: Courtesy of Badger Meter/Research Control Valves.

SAMPLE SYSTEM COMPONENTS

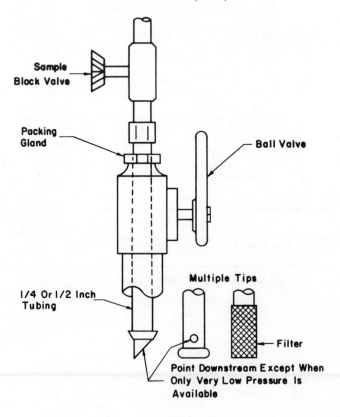

Low Pressure Sample Tap

Sample Block Valve

Packing Gland

Ball Valve

Multiple Tips

1/4 Or 1/2 Inch Tubing

Filter

Point Downstream Except When Only Very Low Pressure Is Available

High Pressure Sample Tap

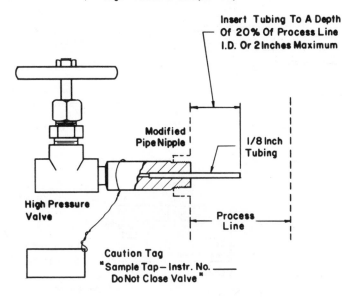

Insert Tubing To A Depth
Of 20% Of Process Line
I.D. Or 2 Inches Maximum

Modified
Pipe Nipple

1/8 Inch
Tubing

High Pressure
Valve

Process
Line

Caution Tag
"Sample Tap – Instr. No. _____
Do Not Close Valve"

Steam Or Electric
Vaporizing Regulator

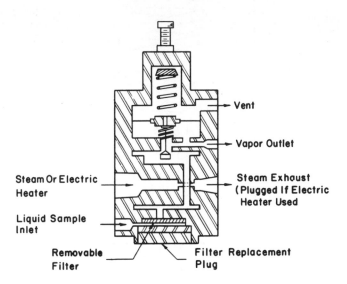

Vent

Vapor Outlet

Steam Or Electric
Heater

Steam Exhaust
(Plugged If Electric
Heater Used

Liquid Sample
Inlet

Removable
Filter

Filter Replacement
Plug

Bypass Filters

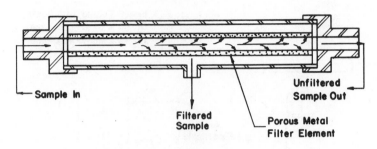

Sample In

Filtered
Sample

Unfiltered
Sample Out

Porous Metal
Filter Element

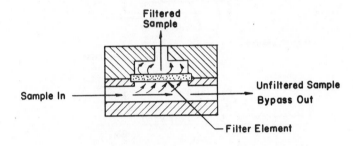

Filtered
Sample

Sample In

Unfiltered Sample
Bypass Out

Filter Element

Condenser / Separator

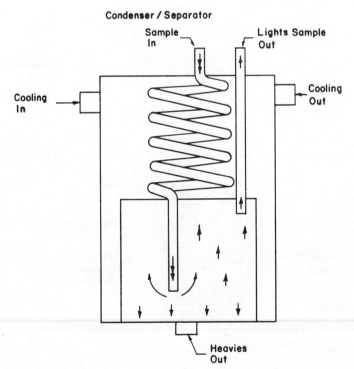

Sample
In

Lights Sample
Out

Cooling
In

Cooling
Out

Heavies
Out

Filter / Separator

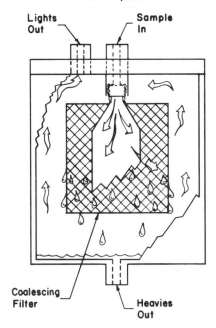

Lights
Out

Sample
In

Coalescing
Filter

Heavies
Out

Filter / Separator

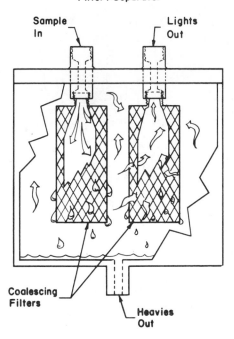

Sample
In

Lights
Out

Coalescing
Filters

Heavies
Out

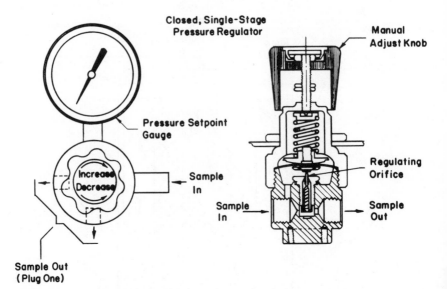

Closed, Single-Stage
Pressure Regulator

Manual
Adjust Knob

Pressure Setpoint
Gauge

Increase
Decrease

Sample
In

Regulating
Orifice

Sample
In

Sample
Out

Sample Out
(Plug One)

Source: Courtesy of GO, Inc.

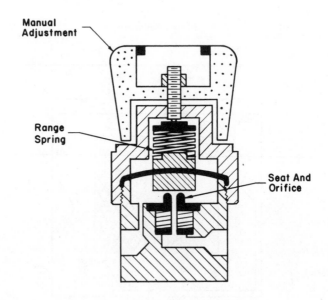

Manual
Adjustment

Range
Spring

Seat And
Orifice

Back Pressure Valve

Source: Courtesy of GO, Inc.

GAS SAFETY DATA

LEGEND

- ■ - Primary Hazard
- R - Recommended
- NR - Not Recommended
- X - Depends on conditions

gas	TOXIC	FLAMMABLE	CORROSIVE	ALUMINUM	COPPER	BRASS	STEEL	STAINLESS STEEL	MONEL	BUTYL	NEOPRENE	POLYETHYLENE	P.V.C.	PENTON	TYGON	KEL-F	TEFLON	SPECIAL CHARACTERISTICS
ACETYLENE		■		R	NR	X	R	R		R	R	R			R	R	R	Do not use at pressures exceeding 15 PSIG.
AIR				R	R	R	R	R		R	R	R	R		R	R	R	
ALLENE		■		R			R	R		R	R	R			R	R	R	
AMMONIA	■			R	NR	NR	X	R		R	R	R	R		R	R	R	Causes stress cracking of copper and copper alloys.
ARGON				R	R	R	R	R		R	R	R	R		R	R	R	
*ARSINE	■			X	NR	X	X	R		R	R	R			R	R	R	Highly Toxic, excessive exposure may have delayed effect.
BORON TRICHLORIDE	■			NR	X	X	X	X	X			X	X	X		R	R	
BORON TRIFLUORIDE	■			X	R	R	R	R	R			X	X	X		R	R	
1-3, BUTADIENE		■		R	R	R	R	R		R	R	R	R		R	R	R	
BUTANE		■		R	R	R	R	R		R	R	R	R		R	R	R	
BUTENES		■		R	R	R	R	R		R	R	R	R		R	R	R	
CARBON DIOXIDE				R	R	R	R	R		R	R	R	R		R	R	R	
CARBON MONOXIDE	■			R	R	R	R	R		R	R	R	R		R	R	R	
CARBONYL SULFIDE	■			R	NR	NR	X	R		R	R	R			R	R	R	Treat as Hydrogen Sulfide, affects central nervous system.
CHLORINE	■			NR	NR	NR	X	X	X	NR	NR	R	X	X	X	R	R	Very toxic and damaging to the respiratory system.
*CYANOGEN	■			X			R	R		X	X	X			R	R	R	Treat as cyanides.
CYCLOPROPANE		■		R	R	R	R	R		R	R	R	R		R	R		
DEUTERIUM		■		R	R	R	R	R		R	R	R	R		R	R		
*DIBORANE	■			R	X	X	R	R		X	X	X			R	R		Highly toxic—high concentrations are pyrophoric.
DIMETHYLAMINE		■		X	NR	NR	R	R		X	X	R	R			X	R	Attacks copper and copper alloys rapidly.
DIMETHYL ETHER		■		R	R	R	R	R		R	R	R				R	R	
ETHANE		■		R	R	R				R	R	R				R	R	
ETHYL ACETYLENE		■		R	R	R				R	R	R				R	R	
ETHYL CHLORIDE		■				R	R	R		R	R	R			R	R	R	
ETHYLENE		■		R	R	R	R	R		R	R	R			R	R	R	
ETHYLENE OXIDE		■			NR	NR	R	R		NR	NR				R	R	R	Exposure of liquid on skin or clothing can cause dermatitis.
*FLUORINE	■			R	R	X		R	R	NR	NR	NR	NR	NR	NR	NR	X	Strong oxidant, can ignite combustible materials and metals.
GERMANE	■			R	R	R	R	R								R	R	
HELIUM				R	R	R	R	R		R	R	R	R		R	R	R	
HEXAFLUOROPROPENE				R	R	R	R	R		R	R	R	R		R	R		
HYDROGEN		■		R	R	R	R	R		R	R	R	R		R	R	R	
HYDROGEN BROMIDE	■			NR	X	NR	X	X	X	X	X	R	R			R	R	Steel or stainless steel serviceable in dry liquid or gas service.
HYDROGEN CHLORIDE	■			NR	X		X	X	X	X	X	R	R			R	R	Steel or stainless steel serviceable in dry liquid or gas service.
*HYDROGEN FLUORIDE	■			X	R	R	R	R				R	R	R		R	R	Exposure can attack skin, bones and fingernails.
*HYDROGEN SELENIDE	■			NR	NR	NR	X	R		X	X	R	R			R	R	Extremely toxic, odor deadens the olfactory nerves.
*HYDROGEN SULFIDE	■			NR	NR	NR	X	R		X	X	R	R			R	R	Odor deadens olfactory nerves, can cause paralysis.

* It is recommended that the user thoroughly familiarize himself with the specific properties of this gas.

Gas safety data courtesy of Scientific Gas Products, Inc.

gas

HAZARDS FOR HUMANS			MATERIALS OF CONSTRUCTION														SPECIAL CHARACTERISTICS
TOXIC	FLAMMABLE	CORROSIVE	ALUMINUM	COPPER	BRASS	STEEL	STAINLESS STEEL	MONEL	BUTYL	NEOPRENE	POLYETHYLENE	P.V.C.	PENTON	TYGON	KEL-F	TEFLON	
ISOBUTANE		■		R	R	R	R	R		R	R	R	R	R	R	R	R	
ISOBUTYLENE		■		R	R	R	R	R		R	R	R	R	R	R	R	R	
KRYPTON				R	R	R	R	R		R	R	R	R	R	R	R	R	
METHANE		■		R	R	R	R	R		R	R	R	R		R	R	R	
METHYL ACETYLENE		■		R	NR	X	R	R		R	R	R	R		R	R	R	
METHYL BROMIDE	■			X	R	R	R	R		R	R	R			R	R		
METHYL CHLORIDE		■		NR	X	R	R	R		R	R	R			R	R		Forms explosive compounds with aluminum.
METHYL MERCAPTAN		■		R	NR	X	R	R		X	X	R			R	R		
MONOETHYLAMINE		■		X	NR	NR	R	R		X	X	R		X	X	R		Attacks copper and copper alloys rapidly.
MONOMETHYLAMINE		■		X	NR	R	R	R		X	X	R		X	X	R		Attacks copper and copper alloys rapidly.
NEON				R	R	R	R	R		R	R	R	R		R	R	R	
*NICKEL CARBONYL	■			R	R	R	R	R			X	R			R	R		Extremely toxic.
NITRIC OXIDE	■		■	R	NR	NR	X	R	NR	X	X	R	R	X	R	R		Readily reacts with Oxygen to form Nitrogen Dioxide
NITROGEN				R	R	R	R	R		R	R	R	R	R	R	R	R	
NITROGEN CHLORIDE	■			X	NR	NR	X	R	NR	NR	NR	R	X		R	R		Extremely toxic and damaging to the respiratory system
NITROGEN TRIOXIDE	■			X	NR	NR	X	R	NR	NR	NR	R	X		R	R		Extremely toxic and damaging to the respiratory system
NITROSYL CHLORIDE	■			NR	NR	NR	NR	NR	R	NR	NR					R		Very corrosive, attacks most metals except nickel.
NITROUS OXIDE				R	R	R	R	R		R	R	R	R		R	R	R	
*OXYGEN			■	R	R	R	R	R		X	X	R	R		R	R	R	Strong oxidant, ignites combustible matter spontaneously
*OZONE	■			R			X	R		NR	NR	X	X				R	Very strong oxidant, irritating to the respiratory system
PHOSGENE	■			NR	NR	NR	X	X	R		X	X	X		R	R		Very toxic.
*PHOSPHINE	■	■		R	X	X	R	R				R	R	R		R	R	Highly toxic—high concentrations are pyrophoric.
PROPANE		■		R	R	R	R	R		R	R	R	R		R	R	R	
PROPYLENE		■		R	R	R	R	R		R	R	R	R		R	R	R	
*SILANE		■		R	R	R	R	R		R	R	R	R		R	R	R	Pyrophoric.
SILICON TETRAFLUORIDE			■	R	R	R	R	R			R					R	R	
SULFUR DIOXIDE	■			R	R	R	R	R			R					R	R	
SULFUR HEXAFLUORIDE				R	R	R	R	R		R	R	R			R	R	R	
SULFUR TETRAFLUORIDE	■			R	R	R	R	R								R	R	
SULFURYL FLUORIDE	■			R	R		R	R			X					R	R	
TETRAFLUOROETHYLENE		■		R	R	R	R	R										
TRIMETHYLAMINE		■		R	NR	NR	R	R		R	X	X	X		X		R	Attacks copper and copper alloys rapidly.
VINYL BROMIDE		■					R	R		R	R	R			R	R	R	
VINYL CHLORIDE		■					R	R		R	R	R			R	R	R	
VINYL FLUORIDE		■					R	R		R	R	R			R	R	R	
XENON				R	R	R	R	R		R	R	R			R	R	R	

LEGEND
■ - Primary Hazard
R - Recommended
NR - Not Recommended
X - Depends on conditions

SGP

* It is recommended that the user thoroughly familiarize himself with the specific properties of this gas.

Gas safety data courtesy of Scientific Gas Products, Inc.

APPENDIX

9

MOISTURE CONVERSION TABLE AND MOISTURE EQUIVALENT VALUE CHART

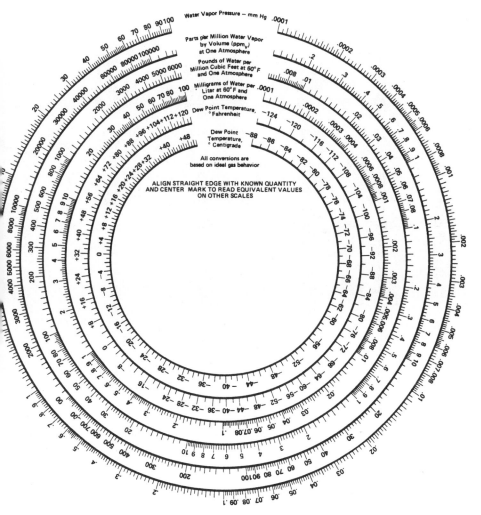

urce: Courtesy of Dr. E. J. Rosa, Ondyne, Inc.

°C	Dewpoint °F	Vapor Pressure (Water/Ice in Equilibrium) mm of Mercury	PPM on Volume Basis at 760 mm of Hg Pressure	Relative Humidity at 70°F%	PPM on Weight Basis in Ai
−110	−166	.0000010	.00132	.0000053	.000
−108	−162	.0000018	.00237	.0000096	.001
−106	−159	.0000028	.00368	.000015	.002
−104	−155	.0000043	.00566	.000023	.003
−102	−152	.0000065	.00855	.000035	.005
−100	−148	.0000099	.0130	.000053	.008
− 98	−144	.000015	.0197	.000080	.012
− 96	−141	.000022	.0289	.00012	.018
− 94	−137	.000033	.0434	.00018	.027
− 92	−134	.000048	.0632	.00026	.039
− 90	−130	.000070	.0921	.00037	.057
− 88	−126	.00010	.132	.00054	.082
− 86	−123	.00014	.184	.00075	.11
− 84	−119	.00020	.263	.00107	.16
− 82	−116	.00029	.382	.00155	.24
− 80	−112	.00040	.562	.00214	.33
− 78	−108	.00056	.737	.00300	.46
− 76	−105	.00077	1.01	.00410	.63
− 74	−101	.00105	1.38	.00559	.86
− 72	− 98	.00143	1.88	.00762	1.17
− 70	− 94	.00194	2.55	.0104	1.58
− 68	− 90	.00261	3.43	.0140	2.13
− 66	− 87	.00349	4.59	.0187	2.84
− 64	− 83	.00464	6.11	.0248	3.79
− 62	− 80	.00614	8.08	.0328	5.01
− 60	− 76	.00808	10.6	.0430	6.59
− 58	− 72	.0106	13.9	.0565	8.63
− 56	− 69	.0138	18.2	.0735	11.3
− 54	− 65	.0178	23.4	.0948	14.5
− 52	− 62	.0230	30.3	.123	18.8
− 50	− 58	.0295	38.8	.157	24.1
− 48	− 54	.0378	49.7	.202	30.9
− 46	− 51	.0481	63.3	.257	39.3
− 44	− 47	.0609	80.0	.325	49.7
− 42	− 44	.0768	101.	.410	62.7
− 40	− 40	.0966	127.	.516	78.9
− 38	− 36	.1209	159.	.644	98.6
− 36	− 33	.1507	198.	.804	122.9
− 34	− 29	.1873	246.	1.00	152.
− 32	− 26	.2318	305.	1.24	189.
− 30	− 22	.2859	376.	1.52	234.
− 28	− 18	.351	462.	1.88	287.
− 26	− 15	.430	566.	2.30	351.
− 24	− 11	.526	692.	2.81	430.
− 22	− 8	.640	842.	3.41	523.
− 20	− 4	.776	1020.	4.13	633.
− 18	− 0	.939	1240.	5.00	770.
− 16	+ 3	1.132	1490.	6.03	925.
− 14	+ 7	1.361	1790.	7.25	1110.
− 12	+ 10	1.632	2150.	8.69	1335.
− 10	+ 14	1.950	2570.	10.4	1596.
− 8	+ 18	2.326	3060.	12.4	1900.
− 6	+ 21	2.765	3640.	14.7	2260.
− 4	+ 25	3.280	4320.	17.5	2680.
− 2	+ 28	3.880	5100.	20.7	3170.
0	+ 32	4.579	6020.	24.4	3640.
+ 2	+ 36	5.294	6970.	28.2	4330.
+ 4	+ 39	6.101	8030.	32.5	4990.
+ 6	+ 43	7.013	9230.	37.4	5730.
+ 8	+ 46	8.045	10590.	42.9	6580.
+ 10	+ 50	9.029	12120.	49.1	7530.
+ 12	+ 54	10.52	13840.	56.1	8600.
+ 14	+ 57	11.99	15780.	63.9	9800.
+ 16	+ 61	13.63	17930.	72.6	11140.
+ 18	+ 64	15.48	20370.	82.5	12650.
+ 20	+ 68	17.54	23080.	93.5	14330.

Source: Courtesy of Dr. E. J. Rosa, Ondyne, Inc.

ABSORBANCE–PERCENT TRANSMITTANCE SCALE

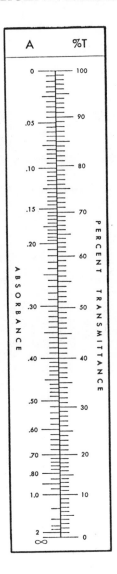

INDEX

Absolute pressure, 29, 47, 186, 223
Absolute temperature, 115, 118
Absorbance, 143
  UV, 234
Absorbance equations, 144, 163
Absorbance range, 143
Absorbance spectrum, 143
Absorptivity, 143–145
Active element, 1, 257
Air entrapment, 84
Airflow characteristics, 22
Alarm, corrosion rate, 112
Alarm relays, 83
Alarm setpoint, conductivity, 80
Ametek/Thermox Company, 115, 116
Amplifier:
  constant-temperature, 229
  electrometer, 229
  high-impedance, 68
AMSCOR, 217
Analytical balance, 204
Analytical Instruments Corp., 219, 230
Analytical laboratory, 6, 193, 210
Analyzer:
  catalytic combustible gas, 60, 127
  chemical oxygen demand, 161
  computer-controlled, 211
  conductivity, 80
  digital, 187, 210
  dissolved oxygen, 8, 114, 129, 131, 135,
    136
  dissolved solids, 81
  dual-beam, 146, 147, 150, 158, 159, 164,
    165
  dual-range, 208
  galvanic, 129, 130, 135, 136
  gasoline octane, 187, 262
  HPLC, 235
  hydrogen in chlorine, 149, 254
  infrared, 19, 60, 137, 143, 150, 151, 155,
    156, 162, 163, 168, 169
  in-line, 6, 262
  laboratory, 262

Larson–Lane condensate, 84
  near IR, 137, 163, 165, 168
  non-contracting IR, 159
  non-dispersive IR, 151
  octane, 207, 210
  on-line, 1, 2, 5, 23, 24, 30, 38, 40, 89,
    207, 236, 239, 241–244, 248, 256
  optical, 11, 27
  oxygen, 114, 122, 127, 128
  paramagnetic, 123, 124, 126, 127
  photometric, 13, 23, 137, 176, 264
  polarographic, 129, 135
  process, 1, 3, 37, 38, 46, 48, 242, 243,
    248, 252
  single-beam, 150, 165
  single-chip, 235
  specific ion, 2
  split-beam, 150
  stack gas, 114, 124, 125
  TC, 151, 161–163
  thermomagnetic, 120, 121, 124, 126, 127
  TOC, 161–163
  total oxygen demand, 161
  trace moisture, 88
  tuned-cavity IR, 157
  ultraviolet, 19, 137, 139, 140, 142, 143,
    146, 148, 150
  UV–VIS, 137, 148, 151, 169, 170, 175
  visible, 143
  wet chemistry, 247
Analyzer audits, annual, 248, 249
Analyzer calibration, 26
Analyzer cell, 2, 13, 20, 32, 33
Analyzer expert, 238, 248, 253
Analyzer house, 8, 12, 14, 19, 26, 27, 39, 47,
  57–60, 257
  multiroom, 6
Analyzer loop, 250
Analyzer obsolescence, 244
Analyzer program, 242
Analyzer shelter, 57
Analyzer specialist, 238
Anarad, Inc., 157

moisture, 10
pH, 7, 10, 53, 71, 76
polyvinyl dichloride, 71
retractable, 111, 266
Ryton®-bodied, 84
stainless steel, 71
submersion, 267
thallium, 114
welded tubing, 110
wire loop, 108
zirconium oxide, 116
Probe aging, 69
Probe fouling, 73
Probe holder:
  flow-through, 72, 84
  universal, 64
Probe resistance, 71
Probe simulator, 70, 110, 264
Process header, 264
Process material, 30, 194
Process samples, 192
Process stream, 1, 5, 27, 37, 40, 66, 68, 127, 132, 139, 172, 173, 184, 220, 264
Process stream components, 264
Process tap, 2, 6, 264
Process variable, 3, 37, 265
Programmable calculators, 39
PROM, 264
Prompting routine, 265
Proportionality constant, 115, 118
Protective equipment, 49
  personal, 50, 52
Purge gas, 24, 265
  nitrogen, 24, 27

Radiation:
  ionizing, 49
  non-ionizing, 49
Radiation protection officer (RPO), 50
Radioactive material, 49, 79
Radioactive sources, 49
Radio-frequency interference (RFI), 73, 82, 255, 265
Radiographic inspection, 50
RAM, 265
Reaction chamber, 172, 208
Reactor:
  glass, 210
  metal, 210
Real-time processing, 255
Recombination effects, 90, 265

Reference time, 231
Reflectance rods, 183
Refractive index (RI), 178, 183–185, 234
Refractometer, 137, 176, 265
  critical-angle, 178, 179, 182–184
Regulator:
  back-pressure, 24, 32, 33, 148
  water, 33
Relief valves, 16, 20, 21, 23, 25, 26, 48, 265
Remote multiplexing, 265
Representative sample, 7, 265
Research octane number (RON), 207, 209
Residence time, 2, 12, 16, 36, 37, 148, 162, 265
Resistivity, 79, 82
Resonant frequency, 200, 201
Response time, 2, 36, 76, 83, 91, 126, 127, 164, 208, 265
Restricting orifice, 8, 12, 27, 125, 198, 265
Retention time, 231, 232, 266
Reverse flow check valve, 266
Reverse osmosis, 84, 85
Rohrback Instruments, 108, 109
Rotameter, 30, 32, 33, 266
  aspirator flow, 34
  bypass flow, 34, 97, 148
  cell flow, 25
  glass tube, 41, 247
  sample, 34
Rotary reaction cell, 188, 190, 195, 199
Routine inspections, 248
Routine maintenance, 242, 266
Rupture disks, 16, 20, 21, 23, 25, 41, 48, 266

Safety, 48
Safety attitude, 48
Safety cable, 10
Safety precautions, 27, 61
Safety regulations, 48, 49
Safety rules, 46, 48, 49
Safety showers, 49, 60, 62
Safety stop, 10
Sample booth, 39, 40, 266
  fiberglass, 39
Sample cell, 24, 25, 27, 34, 163, 165, 266
  Luft-type, 164
  Monel®, 164
  nickel, 164
  stainless steel, 164
Sample cell windows, 12
Sample chamber, 172